PRAISE FOR *Little Brother*

"A believable and frightening tale of a near-future San Francisco... Filled with sharp dialogue and detailed descriptions...within a tautly crafted fictional framework."
   *Publishers Weekly*, STARRED REVIEW

"Readers will delight in the details of how Marcus attempts to stage a techno-revolution... Buy multiple copies; this book will be h4wt (that's 'hot,' for the nonhackers)."
   *Booklist*, STARRED REVIEW

"Marcus is a wonderfully developed character: hyperaware of his surroundings, trying to redress past wrongs, and rebelling against authority... Raising pertinent questions and fostering discussion, this techno-thriller is an outstanding first purchase."
   *School Library Journal*, STARRED REVIEW

"*Little Brother* is generally awesome in the more vernacular sense: It's pretty freaking cool... a fluid, instantly ingratiating fiction writer...he's also terrific at finding the human aura shimmering around technology."
   *Los Angeles Times*

"A wonderful, importa[nt]... I've read this year, an[d]... [wor]lds, male and female, as I[...]
   NEIL GAIMAN, AU[THOR...]

"A worthy younger sib[ling...] is lively, precocious, and most importantly, a little scary."
   BRIAN K. VAUGHAN, AUTHOR OF *Y: The Last Man*

"Scarily realistic... Action-packed with tales of courage, technology, and demonstrations of digital disobed[ience...]
   ANDREW "BU[NNIE]" HUANG, A[UTHOR OF...]

"The right book a[t] the right time from the right author...and, not entirely coincidentally, Cory Doctorow's [b]est novel yet."
   JOHN SCALZ[I...]

"I was completely [h]ooked in the fir[st f]ew minutes."
   MITCH KAPO[R...]
   FRONTIER FO[UNDATION...]

"*Little Brother* is a [...] our country's bes[t] hope for the futu[re...]
   JANE MCGON[IGAL...]

"The teenage voi[ce is pitch-perfect,] [I] [c]ouldn't put it dow[n], and I loved it."
   JO WALTON, [...]

"Doctorow strings together wonderfully witty words into pithy sentences that have no right making as much sense as they do. He brings a powerful but lighthearted magic to a world we very much hope resembles the real world."

*Agony Column*

FOR *Overclocked*:

"*Overclocked: Stories of the Future Present* is really good story telling, good extrapolation on present trends. My sysadmins should check out the first story, 'When Sysadmins Ruled the Earth.'"

CRAIG "CRAIGSLIST" NEWMARK

"In these quirky, brashly engaged 'stories of the future present' Cory Doctorow shows us life from the point-of-view of the plugged-in generation and makes it feel like a totally alien world."

*Montreal Gazette*

"He has a knack for identifying those seminal trends of our current landscape that will in all likelihood determine the shape of our future(s)...."

*Sci Fi Weekly*

"'Overclocked' is a reminder that we can't hope to keep up and shouldn't bother. But we do need to keep alert, to keep ourselves caffeinated, to run as fast as we can — if we hope to stay in the same place. Getting ahead? That's, alas, a thing of the past."

*NPR*

"If you want to glimpse the future of copyright policing, video-game sweatshops, robotic intelligence, info war, and how computer geeks will survive the apocalypse, then this collection of shorts is your oracle.... Doctorow is rapidly emerging as the William Gibson of his generation."

*Entertainment Weekly*

"Each short story is an idea bomb with a candy coating of human drama, wrapped in shiny tech tropes and ready to blow your mind. *Overclocked* is sf info-warfare ammunition of the highest caliber, so load up, move out, and take no prisoners..."

*SFRevu*

"The appealing characters, snappy writing and swift pace will surely tempt the younger and/or geekier sections of the sf audience."

*Kirkus*

> CONTENT

# Content

*Selected Essays on Technology, Creativity, Copyright, and the Future of the Future*

CORY DOCTOROW
*doctorow@craphound.com*

TACHYON PUBLICATIONS | SAN FRANCISCO

8|12
15⁰⁰

Content

Copyright © 2008 by Cory Doctorow
Foreword © 2008 by John Perry Barlow
Cover design © 2008 by Ann Monn
Author photo © 2005 by Patrick H. Lauke aka Redux (*www.splintered.co.uk*)
Interior design and composition by John D. Berry
The typeface is Chaparral Pro, designed by Carol Twombly

Tachyon Publications
1459 18th Street #139
San Francisco, CA 94107
(415) 285-5615
*www.tachyonpublications.com*

Series Editor: Jacob Weisman

ISBN 13: 978-1-892391-81-0
ISBN 10: 1-892391-81-3

Printed in the United States if America by Worzalla

First Edition: 2008

9 8 7 6 5 4 3 2 1

DEDICATION:

For the founders of the Electronic Frontier Foundation:
John Perry Barlow, Mitch Kapor, and John Gilmore

For the staff — past and present — of the Electronic Frontier
Foundation

For the supporters of the Electronic Frontier Foundation

CONTENTS

# Foreword

John Perry Barlow
San Francisco — Seattle — Vancouver — San Francisco
Tuesday, April 1, 2008

"Content," huh? Ha! Where's the container?

Perhaps these words appear to you on the pages of a book, a physical object that might be said to have "contained" the thoughts of my friend and co-conspirator Cory Doctorow as they were transported in boxes and trucks all the way from his marvelous mind into yours. If that is so, I will concede that you *might* be encountering "content." (Actually, if that's the case, I'm delighted on Cory's behalf, since that means that you have also paid him for these thoughts. We still know how to pay creators directly for the works they embed in stuff.)

But the chances are excellent that you're reading these liquid words as bit-states of light on a computer screen, having taken advantage of his willingness to let you have them in that form for free. In such an instance, what "contains" them? Your hard disk? His? The Internet and all the servers and routers in whose caches the ghosts of their passage might still remain? Your mind? Cory's?

To me, it doesn't matter. Even if you're reading this from a book, I'm still not convinced that what you have in your hands is its container, or that, even if we agreed on that point, that a little ink in the shape of, say, the visual pattern you're trained to interpret as meaning "a little ink" in whatever font the publisher chooses, is not, as Magritte would remind us, the same thing as a little ink, even though it is.

Meaning is the issue. If you couldn't read English, this whole book would obviously contain nothing as far as you were concerned. Given that Cory is really cool and interesting, you might be motivated to learn English so that you could read this book, but even then it wouldn't be a container so much as a conduit.

The real "container" would be a process of thought that began when I compressed my notion of what is meant by the word "ink" — which, when it comes to the substances that can be used to make marks on paper, is rather more variable than you might think — and would kind of end when you decompressed it in your own mind as whatever you think it is.

I know this is getting a bit discursive, but I do have a point. Let me just make it so we can move on.

I believe, as I've stated before, that information is simultaneously a relationship, an action, and an area of shared mind. What it isn't is a noun.

Information is not a thing. It isn't an object. It isn't something that, when you sell it or have it stolen, ceases to remain in your possession. It doesn't have a market value that can be objectively determined. It is not, for example, much like a 2004 Ducati ST4S motorcycle, for which I'm presently in the market, and which seems — despite variabilities based on, I must admit, informationally based conditions like mileage and whether it's been dropped — to have a value that is pretty consistent among the specimens I can find for sale on the Web.

Such economic clarity could not be established for anything "in" this book, which you either obtained for free or for whatever price the publisher eventually puts on it. If it's a book you're reading from, then presumably Cory will get paid some percentage of whatever you, or the person who gave it to you, paid for it.

But I won't. I'm not getting paid to write this forward, neither in royalties nor upfront. I am, however, getting some intangible value, as one generally does whenever he does a favor for a friend. For me, the value being retrieved from going to the trouble of writing these words is not so different from the value you retrieve from reading them. We are both mining a deeply intangible "good," which lies in interacting with The Mind of Cory Doctorow. I mention this because it demonstrates the immeasurable role of relationship as the driving force in an information economy.

But neither am I creating content at the moment nor are you "consuming" it (since, unlike a hamburger, these words will remain after you're done with them, and, also unlike a hamburger you won't subsequently, well... never mind.) Unlike real content, like the stuff in a shipping container, these words have neither grams nor liters by which one might measure their value. Unlike gasoline, ten bucks worth of this stuff will get some people a lot further than others, depending on their interest and my eloquence, neither of which can be quantified.

It's this simple: the new meaning of the word "content" is plain wrong. In fact, it is intentionally wrong. It's a usage that only arose when the institutions that had fattened on their ability to bottle and distribute the genius of human expression began to realize that their containers were melting away, along with their reason to be in business. They started calling it content at exactly the time it ceased to be. Previously they had sold books and records and films, all nouns to be sure. They didn't know what to call the mysterious ghosts of thought that were attached to them.

Thus, when not applied to something you can put in a bucket (of whatever size), "content" actually represents a plot to make you think that meaning is a *thing*. It isn't. The only reason they want you to think that it is, is because they know how to own things, how to give them a value based on weight or quantity, and, more to the point, how to make them artificially scarce in order to increase their value.

That, and the fact that after a good twenty-five years of advance warning, they still haven't done much about the Economy of Ideas besides trying to stop it from happening.

As I get older, I become less and less interested in saying "I told you so." But in this case, I find it hard to resist. Back during the Internet equivalent of the Pleistocene, I wrote a piece for an ancestor of *Wired* magazine called *Wired* magazine that was titled, variously, "The Economy of Ideas" or "Wine without Bottles." In this essay, I argued that it would be deucedly difficult to continue to apply the Adam Smithian economic principles regarding the relationship between scarcity and value to any products that could be reproduced and distributed infinitely at zero cost.

I proposed, moreover, that, to the extent that anything might be scarce in such an economy, it would be attention, and that invisibility would be a bad strategy for increasing attention. That, in other words, *familiarity* might convey more value to information than scarcity would.

I did my best to tell the folks in what is now called "The Content Industry" — the institutions that once arose for the useful purpose of conveying creative expression from one mind to many — that this would be a good time to change their economic model. I proposed that copyright had worked largely because it had been difficult, as a practical matter, to make a book or a record or motion picture film spool.

It was my theory that as soon as all human expression could be reduced into ones and zeros, people would begin to realize what this "stuff" really was and come up with an economic paradigm for rewarding its sources that didn't seem as futile as claiming to own the wind. Organizations would adapt. The law would change. The notion of "intellectual property," itself only about thirty-five years old, would be chucked immediately onto the magnificent ash-heap of Civilization's idiotic experiments.

Of course, as we now know, I was wrong. Really wrong.

As is my almost pathological inclination, I extended them too much credit. I imputed to institutions the same capacities for adaptability and recognition of the obvious that I assume for humans. But institutions, having the legal system a fundamental part of their genetic code, are not so readily ductile.

This is particularly true in America, where some combination of certainty and control is the actual "deity" before whose altar we worship, and where we have a regular practice of spawning large and inhuman collective organisms that are a kind of meta-parasite. These critters — let's call them publicly held corporations — may be made out of humans, but they are not human. Given human folly, that characteristic might be semi-OK if they were actually as cold-bloodedly expedient as I once fancied them — yielding only to the will of the markets and the raw self-interest of their shareholders. But no. They are also symbiotically subject to the "religious beliefs" of those humans who feed in their upper elevations.

Unfortunately, the guys (and they mostly are guys) who've been running The Content Industry since it started to die share something like a doctrinal fundamentalism that has led them to such beliefs as the conviction that there's no difference between listening to a song and shop-lifting a toaster.

Moreover, they dwell in such a sublime state of denial that they think they are stewarding the creative process as it arises in the creative humans they exploit savagely — knowing, as they do, that a creative human would rather be heard than paid — and that they, a bunch of sated old scoundrels nearing retirement, would be able to find technological means for wrapping "containers" around "their" "content" that the adolescent electronic Hezbollah they've inspired by suing their own customers will neither be smart nor motivated enough to shred whatever pathetic digital bottles their lackeys design.

And so it has been for the last thirteen years. The companies that claim the ability to regulate humanity's Right to Know have been tireless in their endeavors to prevent the inevitable. They won most of the legislative battles in the U.S. and abroad, having purchased all the government that money could buy. They even won most of the contests in court. They created digital rights management software schemes that behaved rather like computer viruses.

Indeed, they did about everything they could short of seriously examining the actual economics of the situation — it has never been proven to me that illegal downloads are more like shoplifted goods than viral marketing — or trying to come up with a business model that the market might embrace.

Had it been left to the stewardship of the usual suspects, there would scarcely be a word or a note online that you didn't have to pay to experience. There would be increasingly little free speech or any consequence, since free speech is not something anyone can own.

Fortunately there were countervailing forces of all sorts, beginning with the wise folks who designed the Internet in the first place. Then there was something called the Electronic Frontier Foundation which I co-founded, along with Mitch Kapor and John Gilmore, back in 1990. Dedicated to the free exchange of useful information in cyberspace, it seemed at times that I had been right in suggesting then that practically every institution of the Industrial Period would try to crush, or at least own, the Internet. That's a lot of lawyers to have stacked against your cause.

But we had Cory Doctorow.

Had nature not provided us with a Cory Doctorow when we needed one, it would have been necessary for us to invent a time machine and go into the future to fetch another like him. That would be about the only place I can imagine finding such a creature. Cory, as you will learn from his various rants "contained" herein, was perfectly suited to the task of subduing the dinosaurs of content.

He's a little like the guerilla plumber Tuttle in the movie *Brazil*. Armed with a utility belt of improbable gizmos, a wildly overclocked mind, a keyboard he uses like a verbal machine gun, and, best of all, a dark sense of humor, he'd go forth against massive industrial forces and return grinning, if a little beat up.

Indeed, many of the essays collected under this dubious title are not only memoirs of his various campaigns but are themselves the very weapons he used in them. Fortunately, he has spared you some of the more sophisticated utilities he employed. He is not battering you with the nerdy technolingo he commands when stacked up against various minutiacrats, but I assure you that he can speak geek with people who, unlike Cory, think they're being pretty social when they're staring at the other person's shoes.

This was a necessary ability. One of the problems that EFF has to contend with is that even though most of our yet-unborn constituency would agree heartily with our central mission — giving everybody everywhere the right to both address and hear everybody everywhere else — the decisions that will determine the eventual viability of that right are being made now and generally in gatherings invisible to the general public, using terminology, whether technical or legal, that would be the verbal equivalent of chloroform to anyone not conversant with such arcana.

I've often repeated my belief that the first responsibility of a human being is to be a better ancestor. Thus, it seems fitting that the appearance of this book, which details much of Cory's time with the EFF, coincides with the appearance of his first-born child, about whom he is a shameless sentimental gusher.

I would like to think that by the time this newest prodigy,

Poesy Emmeline Fibonacci Nautilus Taylor Doctorow — you see what I mean about paternal enthusiasm — has reached Cory's age of truly advanced adolescence, the world will have recognized that there are better ways to regulate the economy of mind than pretending that its products are something like pig iron. But even if it hasn't, I am certain that the global human discourse will be less encumbered than it would have been had not Cory Doctorow blessed our current little chunk of space/time with his fierce endeavors.

And whatever it is that might be "contained" in the following.

> CONTENT

# Microsoft Research DRM Talk

(Originally given as a talk to Microsoft's Research Group and other interested parties from within the company at their Redmond offices on June 17, 2004.)

Greetings fellow pirates! Arrrrr!

I'm here today to talk to you about copyright, technology, and DRM [digital rights management]. I work for the Electronic Frontier Foundation on copyright stuff (mostly), and I live in London. I'm not a lawyer — I'm a kind of mouthpiece/activist type, though occasionally they shave me and stuff me into my Bar Mitzvah suit and send me to a standards body or the UN to stir up trouble. I spend about three weeks a month on the road doing completely weird stuff like going to Microsoft to talk about DRM.

I lead a double life: I'm also a science fiction writer. That means I've got a dog in this fight, because I've been dreaming of making my living from writing since I was twelve years old. Admittedly, my IP-based biz isn't as big as yours, but I guarantee you that it's every bit as important to me as yours is to you.

Here's what I'm here to convince you of:

1. That DRM systems don't work
2. That DRM systems are bad for society
3. That DRM systems are bad for business
4. That DRM systems are bad for artists
5. That DRM is a bad business-move for MSFT

It's a big brief, this talk. Microsoft has sunk a lot of capital into DRM systems, and spent a lot of time sending folks like Martha and Brian and Peter around to various smoke-filled rooms to make sure that Microsoft DRM finds a hospitable home in the future world. Companies like Microsoft steer like old Buicks, and this issue has a lot of forward momentum that will be hard to soak up without driving the engine block back into the driver's compartment. At best I think that Microsoft might convert some of that momentum on DRM into angular momentum, and in so doing, save all our asses.

Let's dive into it.

## 1. DRM systems don't work

This bit breaks down into two parts:

1. A quick refresher course in crypto theory
2. Applying that to DRM

Cryptography — secret writing — is the practice of keeping secrets. It involves three parties: a sender, a receiver, and an attacker (actually, there can be more attackers, senders and recipients, but let's keep this simple). We usually call these people Alice, Bob, and Carol.

Let's say we're in the days of Caesar, the Gallic War. You need to send messages back and forth to your generals, and you'd prefer that the enemy doesn't get hold of them. You can rely on the idea that anyone who intercepts your message is probably illiterate, but that's a tough bet to stake your empire on. You can put your messages into the hands of reliable messengers who'll chew them up and swallow them if captured — but that doesn't help you if Brad Pitt and his men in skirts skewer him with an arrow before he knows what's hit him.

So you encipher your message with something like ROT-13,

where every character is rotated halfway through the alphabet. They used to do this with non-worksafe material on Usenet, back when anyone on Usenet cared about work-safe-ness — A would become N, B is O, C is P, and so forth. To decipher, you just add 13 more, so N goes to A, O to B, yadda yadda.

Well, this is pretty lame: as soon as anyone figures out your algorithm, your secret is g0nez0red.

So if you're Caesar, you spend a lot of time worrying about keeping the existence of your messengers and their payloads secret. Get that? You're Augustus and you need to send a message to Brad without Caseous (a word I'm reliably informed means "cheese-like, or pertaining to cheese") getting his hands on it. You give the message to Diatomaceous, the fleetest runner in the empire, and you encipher it with ROT-13 and send him out of the garrison in the pitchest hour of the night, making sure no one knows that you've sent it out. Caseous has spies everywhere, in the garrison and staked out on the road, and if one of them puts an arrow through Diatomaceous, they'll have their hands on the message, and then if they figure out the cipher, you're b0rked. So the existence of the message is a secret. The cipher is a secret. The ciphertext is a secret. That's a lot of secrets, and the more secrets you've got, the less secure you are, especially if any of those secrets are shared. Shared secrets aren't really all that secret any longer.

Time passes, stuff happens, and then Tesla invents the radio and Marconi takes credit for it. This is both good news and bad news for crypto: On the one hand, your messages can get to anywhere with a receiver and an antenna, which is great for the brave fifth columnists working behind the enemy lines. On the other hand, anyone with an antenna can listen in on the message, which means that it's no longer practical to keep the existence of the message a secret. Any time Adolf sends a message to Berlin, he can assume Churchill overhears it.

Which is OK, because now we have computers — big, bulky, primitive mechanical computers, but computers still. Computers are machines for rearranging numbers, and so scientists on both sides engage in a fiendish competition to invent the most cleverest method they can for rearranging numerically represented text so that the other side can't unscramble it. The existence of the message isn't a secret anymore, but the cipher is.

But this is still too many secrets. If Bobby intercepts one of Adolf's Enigma machines, he can give Churchill all kinds of intelligence. I mean, this was good news for Churchill and us, but bad news for Adolf. And at the end of the day, it's bad news for anyone who wants to keep a secret.

Enter keys: a cipher that uses a key is still more secure. Even if the cipher is disclosed, even if the ciphertext is intercepted, without the key (or a break), the message is secret. Post-war, this is doubly important as we begin to realize what I think of as Schneier's Law: "Any person can invent a security system so clever that she or he can't think of how to break it." This means that the only experimental methodology for discovering if you've made mistakes in your cipher is to tell all the smart people you can about it and ask them to think of ways to break it. Without this critical step, you'll eventually end up living in a fool's paradise, where your attacker has broken your cipher ages ago and is quietly decrypting all her intercepts of your messages, snickering at you.

Best of all, there's only one secret: the key. And with dual-key crypto it becomes a lot easier for Alice and Bob to keep their keys secret from Carol, even if they've never met. So long as Alice and Bob can keep their keys secret, they can assume that Carol won't gain access to their cleartext messages, even though she has access to the cipher and the ciphertext. Conveniently enough, the keys are the shortest and simplest of the secrets, too: hence even easier to keep away from Carol. Hooray for Bob and Alice.

Now, let's apply this to DRM.

In DRM, the attacker is *also the recipient*. It's not Alice and Bob and Carol, it's just Alice and Bob. Alice sells Bob a DVD. She sells Bob a DVD player. The DVD has a movie on it — say, *Pirates of the Caribbean* — and it's enciphered with an algorithm called CSS — Content Scrambling System. The DVD player has a CSS unscrambler.

Now, let's take stock of what's a secret here: the cipher is well known. The ciphertext is most assuredly in enemy hands, arrr. So what? As long as the key is secret from the attacker, we're golden.

But there's the rub. Alice wants Bob to buy *Pirates of the Caribbean* from her. Bob will only buy *Pirates of the Caribbean* if he can descramble the CSS-encrypted VOB — video object — on his DVD player. Otherwise, the disc is only useful to Bob as a drinks-coaster. So Alice has to provide Bob — the attacker — with the key, the cipher, and the ciphertext.

Hilarity ensues.

DRM systems are usually broken in minutes, sometimes days. Rarely, months. It's not because the people who think them up are stupid. It's not because the people who break them are smart. It's not because there's a flaw in the algorithms. At the end of the day, all DRM systems share a common vulnerability: they provide their attackers with ciphertext, the cipher, and the key. At this point, the secret isn't a secret anymore.

## 2. DRM systems are bad for society

Raise your hand if you're thinking something like, "But DRM doesn't have to be proof against smart attackers, only average individuals! It's like a speedbump!"

Put your hand down.

This is a fallacy for two reasons: one technical, and one social. They're both bad for society, though.

Here's the technical reason: I don't need to be a cracker to break your DRM. I only need to know how to search Google, or Kazaa, or any of the other general-purpose search tools for the cleartext that someone smarter than me has extracted.

Raise your hand if you're thinking something like, "But NGSCB can solve this problem: we'll lock the secrets up on the logic board and goop it all up with epoxy."

Put your hand down.

Raise your hand if you're a co-author of the Darknet paper.

Everyone in the first group, meet the co-authors of the Darknet paper. This is a paper that says, among other things, that DRM will fail for this very reason. Put your hands down, guys.

Here's the social reason that DRM fails: keeping an honest user honest is like keeping a tall user tall. DRM vendors tell us that their technology is meant to be proof against average users, not organized criminal gangs like the Ukrainian pirates who stamp out millions of high-quality counterfeits. It's not meant to be proof against sophisticated college kids. It's not meant to be proof against anyone who knows how to edit her registry, or hold down the shift key at the right moment, or use a search engine. At the end of the day, the user DRM is meant to defend against is the most unsophisticated and least capable among us.

Here's a true story about a user I know who was stopped by DRM. She's smart, college educated, and knows nothing about electronics. She has three kids. She has a DVD in the living room and an old VHS deck in the kids' playroom. One day, she brought home the *Toy Story* DVD for the kids. That's a substantial investment, and given the generally jam-smeared character of everything the kids get their paws on, she decided to tape the DVD off to VHS and give that to the kids — that way she could make a fresh VHS copy when the first one went south. She cabled her DVD into her VHS and pressed play on the DVD and record on the VCR and waited.

Before I go further, I want us all to stop a moment and marvel at this. Here is someone who is practically technophobic, but who was able to construct a mental model of sufficient accuracy that she figured out that she could connect her cables in the right order and dub her digital disc off to analog tape. I imagine that everyone in this room is the front-line tech support for someone in her or his family: Wouldn't it be great if all our non-geek friends and relatives were this clever and imaginative?

I also want to point out that this is the proverbial honest user. She's not making a copy for the next door neighbors. She's not making a copy and selling it on a blanket on Canal Street. She's not ripping it to her hard drive, DivX encoding it, and putting it in her Kazaa sharepoint. She's doing something *honest* — moving it from one format to another. She's home taping.

Except she fails. There's a DRM system called Macrovision embedded — by law — in every VHS that messes with the vertical blanking interval in the signal and causes any tape made in this fashion to fail. Macrovision can be defeated for about $10 with a gadget readily available on eBay. But our infringer doesn't know that. She's "honest." Technically unsophisticated. Not stupid, mind you — just naive.

The Darknet paper addresses this possibility: it even predicts what this person will do in the long run: she'll find out about Kazaa and the next time she wants to get a movie for the kids, she'll download it from the Net and burn it for them.

In order to delay that day for as long as possible, our lawmakers and big rightsholder interests have come up with a disastrous policy called anticircumvention.

Here's how anticircumvention works: if you put a lock — an

access control — around a copyrighted work, it is illegal to break that lock. It's illegal to make a tool that breaks that lock. It's illegal to tell someone how to make that tool. One court even held it illegal to tell someone where she can find out how to make that tool.

Remember Schneier's Law? Anyone can come up with a security system so clever that he can't see its flaws. The only way to find the flaws in security is to disclose the system's workings and invite public feedback. But now we live in a world where any cipher used to fence off a copyrighted work is off-limits to that kind of feedback. That's something that a Princeton engineering prof named Ed Felten and his team discovered when he submitted a paper to an academic conference on the failings in the Secure Digital Music Initiative, a watermarking scheme proposed by the recording industry. The RIAA responded by threatening to sue his ass if he tried it. We fought them because Ed is the kind of client that impact litigators love: unimpeachable and clean-cut and the RIAA folded. Lucky Ed. Maybe the next guy isn't so lucky.

Matter of fact, the next guy wasn't. Dmitry Sklyarov is a Russian programmer who gave a talk at a hacker con in Vegas on the failings in Adobe's ebook locks. The FBI threw him in the slam for thirty days. He copped a plea, went home to Russia, and the Russian equivalent of the State Department issued a blanket warning to its researchers to stay away from American conferences, since we'd apparently turned into the kind of country where certain equations are illegal.

Anticircumvention is a powerful tool for people who want to exclude competitors. If you claim that your car engine firmware is a "copyrighted work," you can sue anyone who makes a tool for interfacing with it. That's not just bad news for mechanics — think of the hotrodders who want to chip their cars to tweak the performance settings. We have companies like Lexmark claiming that their printer cartridges contain copyrighted works — software that trips an "I am empty" flag when the toner runs out, and have sued a competitor who made a remanufactured cartridge that reset the flag. Even garage-door opener companies have gotten in on the act, claiming that their receivers' firmware are copyrighted works. Copyrighted cars, print carts, and garage-door openers: What's next, copyrighted light-fixtures?

Even in the context of legitimate — excuse me, "traditional" — copyrighted works like movies on DVDs, anticircumvention is bad news. Copyright is a delicate balance. It gives creators and their assignees some rights, but it also reserves some rights to the public. For example, an author has no right to prohibit anyone from transcoding his books into assistive formats for the blind. More importantly, though, a creator has a very limited say over what you can do once you lawfully acquire her works. If I buy your book, your painting, or your DVD, it belongs to me. It's my property. Not my "intellectual property" — a whacky kind of pseudo-property that's Swiss-cheesed with exceptions, easements, and limitations — but real, no-fooling, actual tangible *property* — the kind of thing that courts have been managing through property law for centuries.

But anticircumvention lets rightsholders invent new and exciting copyrights for themselves — to write private laws without accountability or deliberation — that expropriate your interest in your physical property to their favor. Region-coded DVDs are an example of this: there's no copyright here or in anywhere I know of that says that an author should be able to control where you enjoy her creative works, once you've paid for them. I can buy a book and throw it in my bag and take it anywhere from Toronto to Timbuktu, and read it wherever I am; I can even buy books in America and bring them to the UK, where the author may have an exclusive distribution deal with a local publisher who sells them for double the U.S. shelf-price. When I'm done with it, I can sell it or give it away in the UK. Copyright lawyers call this "First Sale," but it may be simpler to think of it as "Capitalism."

The keys to decrypt a DVD are controlled by an org called DVD-CCA, and they have a bunch of licensing requirements for anyone who gets a key from them. Among these is something called region-coding: if you buy a DVD in France, it'll have a flag set that says, "I am a European DVD." Bring that DVD to America and your DVD player will compare the flag to its list of permitted regions, and if they don't match, it will tell you that it's not allowed to play your disc.

Remember: there is no copyright that says that an author gets to do this. When we wrote the copyright statutes and granted authors the right to control display, performance, duplication, derivative works, and so forth, we didn't leave out "geography" by accident. That was on purpose.

So when your French DVD won't play in America, that's not because it'd be illegal to do so, it's because the studios have invented a business model and then invented a copyright law to prop it up. The DVD is your property and so is the DVD player, but if you break the region-coding on your disc, you're going to run afoul of anticircumvention.

That's what happened to Jon Johansen, a Norwegian teenager who wanted to watch French DVDs on his Norwegian DVD player. He and some pals wrote some code to break the CSS so that he could do so. He's a wanted man here in America; in Norway the studios put the local fuzz up to bringing him up on charges of *unlawfully trespassing upon a computer system*. When his defense asked, "Which computer has Jon trespassed upon?" the answer was: "His own."

His no-fooling, real and physical property has been expropriated by the weird, notional, metaphorical intellectual property on his DVD: DRM only works if your record player becomes the property of whomever's records you're playing.

## 3. DRM systems are bad for biz

This is the worst of all the ideas embodied by DRM: that people who make record players should be able to spec whose records you can listen to, and that people who make records should have a veto over the design of record players.

We've never had this principle: in fact, we've always had just the reverse. Think about all the things that can be plugged into a parallel or serial interface that were never envisioned by their inventors. Our strong economy and rapid innovation are byproducts of the ability of anyone to make anything that plugs into anything else: from the Flowbee electric razor that snaps onto the end of your vacuum-hose to the octopus spilling out of your car's dashboard lighter socket, standard interfaces that anyone can build for are what makes billionaires out of nerds.

The courts affirm this again and again. It used to be illegal to plug anything that didn't come from AT&T into your phone-jack. They claimed that this was for the safety of the network, but really it was about propping up this little penny-ante racket that AT&T had in charging you a rental fee for your phone until you'd paid for it a thousand times over.

When that ban was struck down, it created the market for third-party phone equipment, from talking novelty phones to answering machines to cordless handsets to headsets — billions of dollars of economic activity that had been suppressed by the closed interface. Note that AT&T was one of the big beneficiaries of this: they *also* got into the business of making phone-kit.

DRM is the software equivalent of these closed hardware interfaces. Robert Scoble is a Softie who has an excellent blog, where he wrote an essay about the best way to protect your investment in the digital music you buy. Should you buy Apple iTunes music or Microsoft DRM music? Scoble argued that Microsoft's music was a sounder investment, because Microsoft would have more downstream licensees for its proprietary format and therefore you'd have a richer ecosystem of devices to choose from when you were shopping for gizmos to play your virtual records on.

What a weird idea: that we should evaluate our record purchases on the basis of which recording company will allow the greatest diversity of record players to play its discs! That's like telling someone to buy the Betamax instead of the Edison Kinetoscope because Thomas Edison is a crank about licensing his patents; all the while ignoring the world's relentless march to the more open VHS format.

It's a bad business. DVD is a format where the guy who makes the records gets to design the record players. Ask yourself: How much innovation has there been over the past decade of DVD players? They've gotten cheaper and smaller, but where are the weird and amazing new markets for DVD that were opened up by the VCR? There's a company that's manufacturing the world's first HDD-based DVD jukebox, a thing that holds 100 movies, and they're charging $27,000 for this thing. We're talking about a few thousand dollars' worth of components — all that other cost is the cost of anticompetition.

## 4. DRM systems are bad for artists

But what of the artist? The hardworking filmmaker, the ink-stained scribbler, the heroin-cured leathery rock-star? We poor slobs of the creative class are everyone's favorite poster-children here: the RIAA and MPAA hold us up and say, "Won't someone please think of the children?" File-sharers say, "Yeah, we're thinking about the artists, but the labels are The Man, who cares what happens to you?"

To understand what DRM does to artists, you need to understand how copyright and technology interact. Copyright is inherently technological, since the things it addresses — copying, transmitting, and so on — are inherently technological.

The piano roll was the first system for cheaply copying music. It was invented at a time when the dominant form of entertainment in America was getting a talented pianist to come into your living room and pound out some tunes while you sang along. The music industry consisted mostly of sheet-music publishers.

The player piano was a digital recording and playback system. Piano-roll companies bought sheet music and ripped the notes printed on it into 0s and 1s on a long roll of computer tape, which they sold by the thousands — the hundreds of thousands — the millions. They did this without a penny's compensation to the publishers. They were digital music pirates. Arrrr!

Predictably, the composers and music publishers went nutso. Sousa showed up in Congress to say that:

> These talking machines are going to ruin the artistic development of music in this country. When I was a boy...in front of every house in the summer evenings, you would find young people together singing the songs of the day or old songs. Today you hear these infernal machines going night and day. We will not have a vocal chord left. The vocal chord will be eliminated by a process of evolution, as was the tail of man when he came from the ape.

The publishers asked Congress to ban the piano roll and to create a law that said that any new system for reproducing music should be subject to a veto from their industry association. Lucky for us,

Congress realized what side of their bread had butter on it and decided not to criminalize the dominant form of entertainment in America.

But there was the problem of paying artists. The Constitution sets out the purpose of American copyright: to promote the useful arts and sciences. The composers had a credible story that they'd do less composing if they weren't paid for it, so Congress needed a fix. Here's what they came up with: anyone who paid a music publisher two cents would have the right to make one piano roll of any song that publisher published. The publisher couldn't say no, and no one had to hire a lawyer at $200 an hour to argue about whether the payment should be two cents or a nickel.

This compulsory license is still in place today: when Joe Cocker sings "With a Little Help from My Friends," he pays a fixed fee to the Beatles' publisher and away he goes — even if Ringo hates the idea. If you ever wondered how Sid Vicious talked Anka into letting him get a crack at "My Way," well, now you know.

That compulsory license created a world where a thousand times more money was made by a thousand times more creators who made a thousand times more music that reached a thousand times more people.

This story repeats itself throughout the technological century, every ten or fifteen years. Radio was enabled by a voluntary blanket license — the music companies got together and asked for a consent decree so that they could offer all their music for a flat fee. Cable TV took a compulsory: the only way cable operators could get their hands on broadcasts was to pirate them and shove them down the wire, and Congress saw fit to legalize this practice rather than screw around with their constituents' TVs.

Sometimes, the courts and Congress decided to simply take away a copyright — that's what happened with the VCR. When Sony brought out the VCR in 1976, the studios had already decided what the experience of watching a movie in your living room would look like: they'd licensed out their programming for use on a machine called a DiscoVision, which played big LP-sized discs that were read-only. Proto-DRM.

The copyright scholars of the day didn't give the VCR very good odds. Sony argued that their box allowed for a fair use, which is defined as a use that a court rules is a defense against infringement based on four factors: whether the use transforms the work into something new, like a collage; whether it uses all or some of the work; whether the work is artistic or mainly factual; and whether the use undercuts the creator's business model.

The Betamax failed on all four fronts: when you time-shifted or duplicated a Hollywood movie off the air, you made a non-transformative use of 100 percent of a creative work in a way that directly undercut the DiscoVision licensing stream.

Jack Valenti, the mouthpiece for the motion-picture industry, told Congress in 1982 that the VCR was to the American film industry "as the Boston Strangler is to a woman home alone."

But the Supreme Court ruled against Hollywood in 1984, when it determined that any device capable of a substantial noninfringing use was legal. In other words, "We don't buy this Boston Strangler business: if your business model can't survive the emergence of this general-purpose tool, it's time to get another business model or go broke."

Hollywood found another business model, as the broadcasters had, as the Vaudeville artists had, as the music publishers had, and they made more art that paid more artists and reached a wider audience.

There's one thing that every new art business model had in common: it embraced the medium it lived in.

This is the overweening characteristic of every single successful new medium: it is true to itself. The Luther Bible didn't suc-

ceed on the axes that made a hand-copied monk Bible valuable:
they were ugly, they weren't in Church Latin, they weren't read
aloud by someone who could interpret it for his lay audience, they
didn't represent years of devoted-with-a-capital-D labor by someone who had given his life over to God. The thing that made the
Luther Bible a success was its scalability: it was more popular because it was more proliferate: all success factors for a new medium
pale beside its profligacy. The most successful organisms on earth
are those that reproduce the most: bugs and bacteria, nematodes
and virii. Reproduction is the best of all survival strategies.

Piano rolls didn't sound as good as the music of a skilled pianist: but they *scaled better*. Radio lacked the social elements of live
performance, but more people could build a crystal set and get it
aimed correctly than could pack into even the largest Vaudeville
house. MP3s don't come with liner notes, they aren't sold to you
by a hipper-than-thou record store clerk who can help you make
your choice, bad rips and truncated files abound: I once downloaded a twelve-second copy of "Hey Jude" from the original Napster. Yet MP3 is outcompeting the CD. I don't know what to do
with CDs anymore: I get them, and they're like the especially nice
garment bag they give you at the fancy suit shop: it's nice and you
feel like a goof for throwing it out, but Christ, how many of these
things can you usefully own? I can put ten thousand songs on
my laptop, but a comparable pile of discs, with liner notes and so
forth — that's a liability: it's a piece of my monthly storage-locker
costs.

Here are the two most important things to know about computers
and the Internet:

1. A computer is a machine for rearranging bits
2. The Internet is a machine for moving bits from one place to
another very cheaply and quickly

Any new medium that takes hold on the Internet and with computers will embrace these two facts, not regret them. A newspaper press is a machine for spitting out cheap and smeary newsprint at speed. If you try to make it output fine art lithos, you'll get junk. If you try to make it output newspapers, you'll get the basis for a free society.

And so it is with the Internet. At the heyday of Napster, record execs used to show up at conferences and tell everyone that Napster was doomed because no one wanted lossily compressed MP3s with no liner notes and truncated files and misspelled metadata.

Today we hear ebook publishers tell each other and anyone who'll listen that the barrier to ebooks is screen resolution. It's bollocks, and so is the whole sermonette about how nice a book looks on your bookcase and how nice it smells and how easy it is to slip into the tub. These are obvious and untrue things, like the idea that radio will catch on once they figure out how to sell you hotdogs during the intermission, or that movies will really hit their stride when we can figure out how to bring the actors out for an encore when the film's run out. Or that what the Protestant Reformation really needs is Luther Bibles with facsimile illumination in the margin and a rent-a-priest to read aloud from your personal Word of God.

New media don't succeed because they're like the old media, only better: they succeed because they're worse than the old media at the stuff the old media is good at, and better at the stuff the old media are bad at. Books are good at being paperwhite, high-resolution, low-infrastructure, cheap and disposable. Ebooks are good at being everywhere in the world at the same time for free in a form that is so malleable that you can just pastebomb it into your IM session or turn it into a page-a-day mailing list.

The only really successful epublishing — I mean, hundreds of thousands, millions of copies distributed and read — is the bookwarez scene, where scanned-and-OCR'd books are distributed on the darknet. The only legit publishers with any success at epub-lishing are the ones whose books cross the Internet without tech-nological fetter: publishers like Baen Books and my own, Tor, who are making some or all of their catalogs available in ASCII and HTML and PDF.

The hardware-dependent ebooks, the DRM use-and-copy-re-stricted ebooks, they're cratering. Sales measured in the tens, sometimes the hundreds. Science fiction is a niche business, but when you're selling copies by the ten, that's not even a business, it's a hobby.

Every one of you has been riding a curve where you read more and more words off of more and more screens every day through most of your professional careers. It's zero-sum: you've also been reading fewer words off of fewer pages as time went by: the dino-sauric executive who prints his email and dictates a reply to his secretary is info-roadkill.

Today, at this very second, people read words off of screens for every hour that they can find. Your kids stare at their Game Boys until their eyes fall out. Euroteens ring doorbells with their hypertrophied, SMS-twitching thumbs instead of their index fingers.

Paper books are the packaging that books come in. Cheap printer-binderies, like the Internet Bookmobile that can produce a full-bleed, four-color, glossy cover, printed spine, perfect-bound book in ten minutes for a dollar are the future of paper books: when you need an instance of a paper book, you generate one, or part of one, and pitch it out when you're done. I landed at SEA-TAC on Monday and burned a couple CDs from my music collec-tion to listen to in the rental car. When I drop the car off, I'll leave them behind. Who needs 'em?

Whenever a new technology has disrupted copyright, we've changed copyright. Copyright isn't an ethical proposition, it's a utilitarian one. There's nothing *moral* about paying a composer tuppence for the piano-roll rights, there's nothing *immoral* about not paying Hollywood for the right to videotape a movie off your TV. They're just the best way of balancing out so that people's physical property rights in their VCRs and phonographs are respected and so that creators get enough of a dangling carrot to go on making shows and music and books and paintings.

Technology that disrupts copyright does so because it simplifies and cheapens creation, reproduction, and distribution. The existing copyright businesses exploit inefficiencies in the old production, reproduction, and distribution system, and they'll be weakened by the new technology. But new technology always gives us more art with a wider reach: that's what tech is *for*.

Tech gives us bigger pies that more artists can get a bite out of. That's been tacitly acknowledged at every stage of the copyfight since the piano roll. When copyright and technology collide, it's copyright that changes.

Which means that today's copyright — the thing that DRM nominally props up — didn't come down off the mountain on two stone tablets. It was created in living memory to accommodate the technical reality created by the inventors of the previous generation. To abandon invention now robs tomorrow's artists of the new businesses and new reach and new audiences that the Internet and the PC can give them.

## 5. DRM is a bad business-move for MSFT

When Sony brought out the VCR, it made a record player that could play Hollywood's records, even if Hollywood didn't like the idea. The industries that grew up on the back of the VCR — movie rentals, home taping, camcorders, even Bar Mitzvah videographers — made billions for Sony and its cohort.

That was good business — even if Sony lost the Betamax-VHS format wars, the money on the world-with-VCRs table was enough to make up for it.

But then Sony acquired a relatively tiny entertainment company and it started to massively screw up. When MP3 rolled around and Sony's Walkman customers were clamoring for a solid-state MP3 player, Sony let its music business-unit run its show: instead of making a high-capacity MP3 Walkman, Sony shipped its Music Clips, low-capacity devices that played brain-damaged DRM formats like Real and OpenMG. They spent good money engineering "features" into these devices that kept their customers from freely moving their music back and forth between their devices. Customers stayed away in droves.

Today, Sony is dead in the water when it comes to Walkmen. The market leaders are poky Singaporean outfits like Creative Labs — the kind of company that Sony used to crush like a bug, back before it got borged by its entertainment unit — and PC companies like Apple.

That's because Sony shipped a product that there was no market demand for. No Sony customer woke up one morning and said, "Damn, I wish Sony would devote some expensive engineering effort in order that I may do less with my music." Presented with an alternative, Sony's customers enthusiastically jumped ship.

The same thing happened to a lot of people I know who used to rip their CDs to WMA. You guys sold them software that produced smaller, better-sounding rips than the MP3 rippers, but you also fixed it so that the songs you ripped were device-locked to their PCs. What that meant is that when they backed up their music to another hard drive and reinstalled their OS (something that the spyware and malware wars has made more common than ever), they discovered that after they restored their music they could no longer play it. The player saw the new OS as a different machine, and locked them out of their own music.

There is no market demand for this "feature." None of your customers want you to make expensive modifications to your products that make backing up and restoring even harder. And there is no moment when your customers will be less forgiving than the moment that they are recovering from catastrophic technology failures.

I speak from experience. Because I buy a new PowerBook every ten months, and because I always order the new models the day they're announced, I get a lot of lemons from Apple. That means that I hit Apple's three-iTunes-authorized-computers limit pretty early on and found myself unable to play the hundreds of dollars' worth of iTunes songs I'd bought because one of my authorized machines was a lemon that Apple had broken up for parts, one was in the shop getting fixed by Apple, and one was my mom's computer, 3,000 miles away in Toronto.

If I had been a less good customer for Apple's hardware, I would have been fine. If I had been a less enthusiastic evangelist for Apple's products — if I hadn't shown my mom how iTunes Music Store worked — I would have been fine. If I hadn't bought so much iTunes music that burning it to CD and re-ripping it and re-keying all my metadata was too daunting a task to consider, I would have been fine.

As it was Apple rewarded my trust, evangelism, and out-of-control spending by treating me like a crook and locking me out of my own music, at a time when my PowerBook was in the shop — i.e., at a time when I was hardly disposed to feel charitable to Apple.

I'm an edge case here, but I'm a *leading edge* case. If Apple succeeds in its business plans, it will only be a matter of time until even average customers have upgraded enough hardware and bought enough music to end up where I am.

You know what I would totally buy? A record player that let me play everybody's records. Right now, the closest I can come to that is an open source app called VLC, but it's clunky and buggy and it didn't come pre-installed on my computer.

Sony didn't make a Betamax that only played the movies that Hollywood was willing to permit — Hollywood asked them to do it, they proposed an early, analog broadcast flag that VCRs could hunt for and respond to by disabling recording. Sony ignored them and made the product they thought their customers wanted.

I'm a Microsoft customer. Like millions of other Microsoft customers, I want a player that plays anything I throw at it, and I think that you are just the company to give it to me.

Yes, this would violate copyright law as it stands, but Microsoft has been making tools of piracy that change copyright law for decades now. Outlook, Exchange, and MSN are tools that abet widescale digital infringement.

More significantly, IIS and your caching proxies all make and serve copies of documents without their authors' consent, something that, if it is legal today, is only legal because companies like Microsoft went ahead and did it and dared lawmakers to prosecute.

Microsoft stood up for its customers and for progress, and won so decisively that most people never even realized that there was a fight.

Do it again! This is a company that looks the world's roughest, toughest anti-trust regulators in the eye and laughs. Compared to anti-trust people, copyright lawmakers are pantywaists. You can take them with your arm behind your back.

In Siva Vaidhyanathan's book *The Anarchist in the Library*, he talks about why the studios are so blind to their customers' desires. It's because people like you and me spent the '80s and the '90s telling them bad science fiction stories about impossible DRM technology that would let them charge a small sum of money every time someone looked at a movie — want to fast-forward? That feature costs another penny. Pausing is two cents an hour. The mute button will cost you a quarter.

When Mako Analysis issued their report last month advising phone companies to stop supporting Symbian phones, they were just writing the latest installment in this story. Mako says that phones like my P900, which can play MP3s as ringtones, are bad for the cellphone economy because it'll put the extortionate ringtone sellers out of business. What Mako is saying is that just because you bought the CD doesn't mean that you should expect to have the ability to listen to it on your MP3 player, and just because it plays on your MP3 player is no reason to expect it to run as a ringtone. I wonder how they feel about alarm clocks that will play a CD to wake you up in the morning? Is that strangling the nascent "alarm tone" market?

The phone companies' customers want Symbian phones and for now, at least, the phone companies understand that if they don't sell them, someone else will.

The market opportunity for truly capable devices is enormous. There's a company out there charging $27,000 for a DVD jukebox — go and eat their lunch! Steve Jobs isn't going to do it: he's off at the D conference telling studio execs not to release hi-def movies until they're sure no one will make a hi-def DVD burner that works with a PC.

Maybe they won't buy into his BS, but they're also not much interested in what you have to sell. At the Broadcast Protection Discussion Group meetings where the Broadcast Flag was hammered out, the studios' position was, "We'll take anyone's DRM except Microsoft's and Philips'." When I met with UK broadcast wonks about the European version of the Broadcast Flag underway at the Digital Video Broadcasters' forum, they told me, "Well, it's different in Europe: mostly they're worried that some American company like Microsoft will get their claws into European television."

American film studios didn't want the Japanese electronics companies to get a piece of the movie pie, so they fought the VCR. Today, everyone who makes movies agrees that they don't want to let you guys get between them and their customers.

Sony didn't get permission. Neither should you. Go build the record player that can play everyone's records.

Because if you don't do it, someone else will.

# The DRM Sausage Factory

(Originally published as "A Behind-the-Scenes Look at How DRM Becomes Law," *InformationWeek*, July 11, 2007.)

Otto von Bismarck quipped, "Laws are like sausages, it is better not to see them being made." I've seen sausages made. I've seen laws made. Both pale in comparison to the process by which anti-copying technology agreements are made.

This technology, usually called "Digital Rights Management" (DRM), proposes to make your computer worse at copying some of the files on its hard drive or on other media. Since all computer operations involve copying, this is a daunting task — as security expert Bruce Schneier has said, "Making bits harder to copy is like making water that's less wet."

At root, DRMs are technologies that treat the owner of a computer or other device as an attacker, someone against whom the system must be armored. Like the electrical meter on the side of your house, a DRM is a technology that you possess, but that you are never supposed to be able to manipulate or modify. Unlike your meter, though, a DRM that is defeated in one place is defeated in all places, nearly simultaneously. That is to say, once someone takes the DRM off a song or movie or ebook, that freed collection of bits can be sent to anyone else, anywhere the network reaches, in an eyeblink. DRM crackers need cunning; those who receive the fruits of their labor need only know how to download files from the Internet.

Why manufacture a device that attacks its owner? A priori, one would assume that such a device would cost more to make than a friendlier one, and that customers would prefer not to buy devices that treat them as presumptive criminals. DRM technologies limit more than copying: they limit ranges of uses, such as viewing a movie in a different country, copying a song to a different manufacturer's player, or even pausing a movie for too long. Surely, this stuff hurts sales: Who goes into a store and asks, "Do you have any music that's locked to just one company's player? I'm in the market for some lock-in."

So why do manufacturers do it? As with many strange behaviors, there's a carrot at play here, and a stick.

The carrot is the entertainment industries' promise of access to their copyrighted works. Add DRM to your iPhone and we'll supply music for it. Add DRM to your TiVo and we'll let you plug it into our satellite receivers. Add DRM to your Zune and we'll let you retail our music in your Zune store.

The stick is the entertainment industries' threat of lawsuits for companies that don't comply. In the last century, entertainment companies fought over the creation of records, radios, jukeboxes, cable TV, VCRs, MP3 players, and other technologies that made it possible to experience a copyrighted work in a new way without permission. There's one battle that serves as the archetype for the rest: the fight over the VCR.

The film studios were outraged by Sony's creation of the VCR. They had found a DRM supplier they preferred, a company called DiscoVision that made non-recordable optical discs. DiscoVision was the only company authorized to play back movies in your living room. The only way to get a copyrighted work onto a VCR cassette was to record it off the TV, without permission. The studios argued that Sony — whose Betamax was the canary in this legal coalmine — was breaking the law by unjustly endangering their revenue from DiscoVision royalties. Sure, they *could* just sell pre-recorded Betamax tapes, but Betamax was a read-write medium: they could be *copied*. Moreover, your personal library of Betamax recordings of the Sunday night movie would eat into the market for DiscoVision discs: Why would anyone buy a pre-recorded video cassette when they could amass all the video they needed with a home recorder and a set of rabbit-ears?

The Supreme Court threw out these arguments in a 1984 5-4 decision, the "Betamax Decision." This decision held that the VCR was legal because it was "capable of sustaining a substantially non-infringing use." That means that if you make a technology that your customers *can* use legally, you're not on the hook for the illegal stuff they do.

This principle guided the creation of virtually every piece of IT invented since: the Web, search engines, YouTube, Blogger, Skype, ICQ, AOL, MySpace... You name it, if it's possible to violate copyright with it, the thing that made it possible is the Betamax principle.

Unfortunately, the Supremes shot the Betamax principle in the gut two years ago, with the Grokster decision. This decision says that a company can be found liable for its customers' bad acts if they can be shown to have "induced" copyright infringement. So, if your company advertises your product for an infringing use, or if it can be shown that you had infringement in mind at the design stage, you can be found liable for your customers' copying. The studios and record labels and broadcasters *love* this ruling, and they like to think that it's even broader than what the courts set out. For example, Viacom is suing Google for inducing copyright infringement by allowing YouTube users to flag some of their videos as private. Private videos can't be found by Viacom's copyright-enforcement bots, so Viacom says that privacy should be illegal, and that companies that give you the option of privacy should be sued for anything you do behind closed doors.

The gutshot Betamax doctrine will bleed out all over the industry for decades (or until the courts or Congress restore it to health), providing a grisly reminder of what happens to companies that try to pour the entertainment companies' old wine into new digital bottles without permission. The tape-recorder was legal, but the digital tape-recorder is an inducement to infringement, and must be stopped.

The promise of access to content and the threat of legal execution for non-compliance is enough to lure technology's biggest players to the DRM table.

I started attending DRM meetings in March 2002, on behalf of my former employers, the Electronic Frontier Foundation. My first meeting was the one where the Broadcast Flag was born. The Broadcast Flag was weird even by DRM standards. Broadcasters are required, by law, to deliver TV and radio without DRM, so that any standards-compliant receiver can receive them. The airwaves belong to the public, and are loaned to broadcasters who have to promise to serve the public interest in exchange. But the MPAA and the broadcasters wanted to add DRM to digital TV, and so they proposed that a law should be passed that would make all manufacturers promise to *pretend* that there was DRM on broadcast signals, receiving them and immediately squirreling them away in encrypted form.

The Broadcast Flag was hammered out in a group called the Broadcast Protection Discussion Group (BPDG), a sub-group from the MPAA's "Content Protection Technology Working Group," which also included reps from all the big IT companies (Microsoft, Apple, Intel, and so on), consumer electronics companies (Panasonic, Philips, Zenith), cable companies, satellite companies, and anyone else who wanted to pay $100 to attend the "public" meetings, held every six weeks or so. (You can attend these meetings yourself if you find yourself near LAX on one of the upcoming dates.)

CPTWG (pronounced Cee-Pee-Twig) is a venerable presence in the DRM world. It was at CPTWG that the DRM for DVDs was hammered out. CPTWG meetings open with a "benediction," delivered by a lawyer, who reminds everyone there that what they say might be quoted "on the front page of the *New York Times*" (though journalists are barred from attending CPTWG meetings and no minutes are published by the organization), and reminding all present not to do anything that would raise eyebrows at the FTC's anti-trust division (I could swear I've seen the Microsoft people giggling during this part, though that may have been my imagination).

The first part of the meeting is usually taken up with administrative business and presentations from DRM vendors, who promise that this time they've really, really figured out how to make computers worse at copying. The real meat comes after the lunch, when the group splits into a series of smaller meetings, many of them closed-door and private (the representatives of the organizations responsible for managing DRM on DVDs splinter off at this point).

Then comes the working group meetings, like the BPDG. The BPDG was nominally set up to create the rules for the Broadcast Flag. Under the Flag, manufacturers would be required to limit their "outputs and recording methods" to a set of "approved technologies." Naturally, every manufacturer in the room showed up with a technology to add to the list of approved technologies — and the sneakier ones showed up with reasons why their competitors' technologies *shouldn't* be approved. If the Broadcast Flag became law, a spot on the "approved technologies" list would be a license to print money: everyone who built a next-gen digital TV would be required, by law, to buy only approved technologies for their gear.

The CPTWG determined that there would be three "chairmen"

of the meetings: a representative from the broadcasters, a representative from the studios, and a representative from the IT industry (note that no "consumer rights" chair was contemplated — we proposed one and got laughed off the agenda). The IT chair was filled by an Intel representative, who seemed pleased that the MPAA chair, Fox Studios's Andy Setos, began the process by proposing that the approved technologies should include only two technologies, both of which Intel partially owned.

Intel's presence on the committee was both reassurance and threat: Reassurance because Intel signaled the fundamental reasonableness of the MPAA's requirements — why would a company with a bigger turnover than the whole movie industry show up if the negotiations weren't worth having? Threat because Intel was poised to gain an advantage that might be denied to its competitors.

We settled in for a long negotiation. The discussions were drawn out and heated. At regular intervals, the MPAA reps told us that we were wasting time — if we didn't hurry things along, the world would move on and consumers would grow accustomed to un-crippled digital TVs. Moreover, Representative Billy Tauzin, the lawmaker who'd evidently promised to enact the Broadcast Flag into law, was growing impatient. The warnings were delivered in quackspeak, urgent and crackling, whenever the discussions dragged, like the crack of the commissars' pistols, urging us forward.

You'd think that a "technology working group" would concern itself with technology, but there was precious little discussion of bits and bytes, ciphers and keys. Instead, we focused on what amounted to contractual terms: If your technology got approved as a DTV "output," what obligations would you have to assume? If a TiVo could serve as an "output" for a receiver, what outputs would the TiVo be allowed to have?

The longer we sat there, the more snarled these contractual terms became: winning a coveted spot on the "approved technology" list would be quite a burden! Once you were in the club, there were all sorts of rules about whom you could associate with, how you had to comport yourself, and so on.

One of these rules of conduct was "robustness." As a condition of approval, manufacturers would have to harden their technologies so that their customers wouldn't be able to modify, improve upon, or even understand their workings. As you might imagine, the people who made open source TV tuners were not thrilled about this, as "open source" and "non-user-modifiable" are polar opposites.

Another was "renewability": the ability of the studios to revoke outputs that had been compromised in the field. The studios expected the manufacturers to make products with remote "kill switches" that could be used to shut down part or all of their device if someone, somewhere had figured out how to do something naughty with it. They promised that we'd establish criteria for renewability later, and that it would all be "fair."

But we soldiered on. The MPAA had a gift for resolving the worst snarls: when shouting failed, they'd lead any recalcitrant player out of the room and negotiate in secret with them, leaving the rest of us to cool our heels. Once, they took the Microsoft team out of the room for *six hours,* then came back and announced that digital video would be allowed to output on non-DRM monitors at a greatly reduced resolution (this "feature" appears in Vista as "fuzzing").

The further we went, the more nervous everyone became. We were headed for the real meat of the negotiations: the *criteria* by which approved technology would be evaluated: How many bits of crypto would you need? Which ciphers would be permissible? Which features would and wouldn't be allowed?

Then the MPAA dropped the other shoe: the sole criteria for inclusion on the list would be the approval of one of its member-companies, or a quorum of broadcasters. In other words, the Broadcast Flag wouldn't be an "objective standard," describing the technical means by which video would be locked away — it would be purely subjective, up to the whim of the studios. You could have the best product in the world, and they wouldn't approve it if your business-development guys hadn't bought enough drinks for their business-development guys at a CES party.

To add insult to injury, the only technologies that the MPAA were willing to consider for initial inclusion as "approved" were the two that Intel was involved with. The Intel co-chairman had a hard time hiding his grin. He'd acted as Judas goat, luring in Apple, Microsoft, and the rest, to legitimize a process that would force them to license Intel's patents for every TV technology they shipped until the end of time.

Why did the MPAA give Intel such a sweetheart deal? At the time, I figured that this was just straight quid pro quo, like Hannibal said to Clarice. But over the years, I started to see a larger pattern: Hollywood likes DRM consortia, and they hate individual DRM vendors. (I've written an entire article about this, but here's the gist: a single vendor who succeeds can name their price and terms — think of Apple or Macrovision — while a consortium is a more easily divided rabble, susceptible to co-option in order to produce ever-worsening technologies — think of Blu-ray and HD-DVD.) Intel's technologies were held through two consortia, the 5C and 4C groups.

The single-vendor manufacturers were livid at being locked out of the digital TV market. The final report of the consortium reflected this — a few sheets written by the chairmen describing the "consensus" and hundreds of pages of angry invective from manufacturers and consumer groups decrying it as a sham.

Tauzin washed his hands of the process: a canny, sleazy Hill operator, he had the political instincts to get his name off any proposal that could be shown to be a plot to break voters' televisions. (Tauzin found a better industry to shill for, the pharmaceutical firms, who rewarded him with a $2,000,000/year job as chief of PHARMA, the pharmaceutical lobby.)

Even Representative Ernest "Fritz" Hollings ("The Senator from Disney," who once proposed a bill requiring entertainment industry oversight of all technologies capable of copying) backed away from proposing a bill that would turn the Broadcast Flag into law. Instead, Hollings sent a memo to Michael Powell, then-head of the FCC, telling him that the FCC already had jurisdiction to enact a Broadcast Flag regulation, without Congressional oversight.

Powell's staff put Hollings's letter online, as they are required to do by federal sunshine laws. The memo arrived as a Microsoft Word file — which EFF then downloaded and analyzed. Word stashes the identity of a document's author in the file metadata, which is how EFF discovered that the document had been written by a staffer at the MPAA.

This was truly remarkable. Hollings was a powerful committee chairman, one who had taken immense sums of money from the industries he was supposed to be regulating. It's easy to be cynical about this kind of thing, but it's genuinely unforgivable: politicians draw a public salary to sit in public office and work for the public good. They're supposed to be working for us, not their donors.

But we all know that this isn't true. Politicians are happy to give special favors to their pals in industry. However, the Hollings memo was beyond the pale. Staffers for the MPAA were writing Hollings's memos, memos that Hollings then signed and mailed off to the heads of major governmental agencies.

The best part was that the legal eagles at the MPAA were wrong. On the advice of "Hollings," the FCC enacted a Broadcast Flag regulation that was almost identical to the proposal from the BPDG, turning themselves into America's "device czars," able to burden any digital technology with "robustness," "compliance," and "revocation rules." The rule lasted just long enough for the DC Circuit Court of Appeals to strike it down and slap the FCC for grabbing unprecedented jurisdiction over the devices in our living rooms.

So ended the saga of the Broadcast Flag. More or less. In the years since the Flag was proposed, there have been several attempts to reintroduce it through legislation, all failed. And as more and more innovative, open devices like the Neuros OSD enter the market, it gets harder and harder to imagine that Americans will accept a mandate that takes away all that functionality.

But the spirit of the Broadcast Flag lives on. DRM consortia are all the rage now — outfits like AACS LA, the folks who control the DRM in Blu-ray and HD-DVD, are thriving and making headlines by issuing fatwas against people who publish their secret integers. In Europe, a DRM consortium working under the auspices of the Digital Video Broadcasters Forum (DVB) has just shipped a proposed standard for digital TV DRM that makes the Broadcast Flag look like the work of patchouli-scented infohippies. The DVB proposal would give DRM consortium the ability to define what is and isn't a valid "household" for the purposes of sharing your video within your "household's devices." It limits how long you're allowed to pause a video, and allows for restrictions to be put in place for hundreds of years, longer than any copyright system in the world would protect any work for.

If all this stuff seems a little sneaky, underhanded, and even illegal to you, you're not alone. When representatives of nearly all the world's entertainment, technology, broadcast, satellite, and cable companies gather in a room to collude to cripple their offerings, limit their innovation, and restrict the market, regulators take notice.

That's why the EU is taking a hard look at HD-DVD and Blu-ray. These systems aren't designed: they're *governed*, and the governors are a shadowy group of offshore giants who answer to no one — not even their own members! I once called the DVD-Copy Control Association (DVD-CCA) on behalf of a Time-Warner magazine, *Popular Science,* for a comment about their DRM. Not only wouldn't they allow me to speak to a spokesman, the person who denied my request also refused to be identified.

The sausage factory grinds away, but today, more activists than ever are finding ways to participate in the negotiations, slowing them up, making them account for themselves to the public. And so long as you, the technology-buying public, pay attention to what's going on, the activists will continue to hold back the tide.

# Happy Meal Toys versus Copyright: How America Chose Hollywood and Wal-Mart, and Why It's Doomed Us, and How We Might Survive Anyway

(Originally published as "How Hollywood, Congress, and DRM Are Beating Up the American Economy," *InformationWeek*, June 11, 2007.)

Back in 1985, the Senate was ready to clobber the music industry for exposing America's impressionable youngsters to sex, drugs, and rock-and-roll. Today, the Attorney General is proposing to give the RIAA legal tools to attack people who attempt infringement.

Through most of America's history, the U.S. government has been at odds with the entertainment giants, treating them as purveyors of filth. But not anymore: today, the U.S. Trade Rep is using America's political clout to force Russia to institute police inspections of its CD presses. (Savor the irony: post-Soviet Russia forgoes its hard-won freedom of the press to protect Disney and Universal!)

How did entertainment go from trenchcoat pervert to top trade priority? I blame the "Information Economy."

No one really knows what "Information Economy" means, but by the early '90s, we knew it was coming. America deployed her least reliable strategic resource to puzzle out what an "information economy" was and to figure out how to ensure America stayed atop the "new economy" — America sent in the futurists.

We make the future in much the same way as we make the

past. We don't remember everything that happened to us, just selective details. We weave our memories together on demand, filling in any empty spaces with the present, which is lying around in great abundance. In *Stumbling on Happiness,* Harvard psych prof Daniel Gilbert describes an experiment in which people with delicious lunches in front of them are asked to remember their breakfast: overwhelmingly, the people with good lunches have more positive memories of breakfast than those who have bad lunches. We don't remember breakfast — we look at lunch and superimpose it on breakfast.

We make the future in the same way: we extrapolate as much as we can, and whenever we run out of imagination, we just shovel the present into the holes. That's why our pictures of the future always seem to resemble the present, only more so.

So the futurists told us about the Information Economy: they took all the "information-based" businesses (music, movies, and microcode, in the neat coinage of Neal Stephenson's 1992 novel *Snow Crash*) and projected a future in which these would grow to dominate the world's economies.

There was only one fly in the ointment: most of the world's economies consist of poor people who have more time than money, and if there's any lesson to learn from American college kids, it's that people with more time than money would rather copy information than pay for it.

Of course they would! Why, when America was aborning, she was a pirate nation, cheerfully copying the inventions of European authors and inventors. Why not? The fledgling revolutionary republic could copy without paying, keep the money on her shores, and enrich herself with the products and ideas of imperial Europe. Of course, once the U.S. became a global hitter in the creative industries, out came the international copyright agreements: the U.S. signed agreements to protect British authors in exchange for reciprocal agreements from the Brits to protect American authors.

It's hard to see why a developing country would opt to export its GDP to a rich country when it could get the same benefit by mere copying. The U.S. would have to sweeten the pot.

The pot-sweetener is the elimination of international trade-barriers. Historically, the U.S. has used tariffs to limit the import of manufactured goods from abroad, and to encourage the import of raw materials from abroad. Generally speaking, rich countries import poor countries' raw materials, process them into manu-factured goods, and export them again. Globally speaking, if your country imports sugar and exports sugar cane, chances are you're poor. If your country imports wood and sells paper, chances are you're rich.

In 1995, the U.S. signed onto the World Trade Organization and its associated copyright and patent agreement, the TRIPS Agreement, and the American economy was transformed.

Any fellow signatory to the WTO/TRIPS can export manu-factured goods to the U.S.A. without any tariffs. If it costs you $5 to manufacture and ship a plastic bucket from your factory in Shenjin Province to the U.S.A., you can sell it for $6 and turn a $1 profit. And if it costs an American manufacturer $10 to make the same bucket, the American manufacturer is out of luck.

The kicker is this: if you want to export your finished goods to America, you have to sign up to protect American copyrights in your own country. Quid pro quo.

The practical upshot, twelve years later, is that most American manufacturing has gone belly up, Wal-Mart is filled with Happy Meal toys and other cheaply manufactured plastic goods, and the whole world has signed onto U.S. copyright laws.

But signing onto those laws doesn't mean you'll enforce them. Sure, where a country is really over a barrel (cough, Russia, cough), they'll take the occasional *pro forma* step to enforce U.S. copyrights, no matter how ridiculous and totalitarian it makes them appear. But with the monthly Russian per-capita GDP hovering at $200, it's just not plausible that Russians are going to start paying $15 for a CD, nor is it likely that they'll stop listening to music until their economy picks up.

But the real action is in China, where pressing bootleg media is a national sport. China keeps promising that it will do something about this, but it's not like the U.S. has any recourse if China drags its heels. Trade courts may find against China, but China holds all the cards. The U.S. can't afford to abandon Chinese manufacturing (and no one will vote for the politician who hextuples the cost of Wi-Fi cards, brassieres, iPods, staplers, yoga mats, and spatulas by cutting off trade with China). The Chinese can just sit tight.

The futurists were just plain wrong. An "information economy" can't be based on selling information. Information technology makes copying information easier and easier. The more IT you have, the less control you have over the bits you send out into the world. It will never, ever, EVER get any harder to copy information from here on in. The information economy is about selling everything *except* information.

The U.S. traded its manufacturing sector's health for its entertainment industry, hoping that *Police Academy* sequels could take the place of the rustbelt. The U.S. bet wrong.

But like a losing gambler who keeps on doubling down, the U.S. doesn't know when to quit. It keeps meeting with its entertainment giants, asking how U.S. foreign and domestic policy can preserve its business model. Criminalize 70 million American file-sharers? Check. Turn the world's copyright laws upside down? Check. Cream the IT industry by criminalizing attempted infringement? Check.

It'll never work. It can never work. There will always be an entertainment industry, but not one based on excluding access to published digital works. Once it's in the world, it'll be copied. This is why I give away digital copies of my books and make money on the printed editions: I'm not going to stop people from copying the electronic editions, so I might as well treat them as an entice-ment to buy the printed objects.

But there *is* an information economy. You don't even need a computer to participate. My barber, an avowed technophobe who rebuilds antique motorcycles and doesn't own a PC, benefited from the information economy when I found him by googling for barbershops in my neighborhood.

Teachers benefit from the information economy when they share lesson plans with their colleagues around the world by email. Doctors benefit from the information economy when they move their patient files to efficient digital formats. Insurance compa-nies benefit from the information economy through better access to fresh data used in the preparation of actuarial tables. Marinas benefit from the information economy when office-slaves look up the weekend's weather online and decide to skip out on Friday for a weekend's sailing. Families of migrant workers benefit from the information economy when their sons and daughters wire cash home from a convenience store's Western Union terminal.

This stuff generates wealth for those who practice it. It en-riches the country and improves our lives.

And it *can* peacefully co-exist with movies, music, and micro-code, but not if Hollywood gets to call the shots. Where IT man-agers are expected to police their networks and systems for un-authorized copying — no matter what that does to productivity — they cannot co-exist. Where our operating systems are rendered inoperable by "copy protection," they cannot co-exist. Where our educational institutions are turned into conscript enforcers for the record industry, they cannot co-exist.

The information economy is all around us. The countries that embrace it will emerge as global economic superpowers. The countries that stubbornly hold to the simplistic idea that the information economy is about selling information will end up at the bottom of the pile.

What country do you want to live in?

# Why Is Hollywood Making a Sequel to the Napster Wars?

(Originally published in *InformationWeek*, August 14, 2007.)

Hollywood loves sequels — they're generally a safe bet, provided that you're continuing an already successful franchise. But you'd have to be nuts to shoot a sequel to a disastrous flop — say, *The Adventures of Pluto Nash* or *Town and Country*.

As disastrous as *Pluto Nash* was, it was practically painless when compared to the Napster debacle. That shipwreck took place six years ago, when the record industry succeeded in shutting down the pioneering file-sharing service, and they show no signs of recovery.

*The disastrous thing about Napster wasn't that it existed, but rather that the record industry managed to kill it.*

Napster had an industry-friendly business model: raise venture capital, start charging for access to the service, and then pay billions of dollars to the record companies in exchange for licenses to their works. Yes, they kicked this plan off without getting permission from the record companies, but that's not so unusual. The record companies followed the same business plan a hundred years ago, when they started recording sheet music without permission, raising capital and garnering profits, and *then* working out a deal to pay the composers for the works they'd built their fortunes on.

Napster's plan was plausible. They had the fastest-adopted technology in the history of the world, garnering 52,000,000 users in eighteen months — more than had voted for either candidate in the preceding U.S. election! — and discovering, via surveys, that a sizable portion would happily pay between $10 and $15 a month for the service. What's more, Napster's architecture included a gatekeeper that could be used to lock-out non-paying users.

The record industry refused to deal. Instead, they sued, bringing Napster to its knees. Bertelsmann bought Napster out of the ensuing bankruptcy, a pattern that was followed by other music giants, like Universal, who slayed MP3.com in the courts, then brought home the corpse on the cheap, running it as an internal project.

After that, the record companies had a field day: practically every venture-funded P2P company went down, and millions of dollars were funneled from the tech venture capital firms on Sand Hill Road to the RIAA's members, using P2P companies and the courts as conduits.

But the record companies weren't ready to replace these services with equally compelling alternatives. Instead, they fielded inferior replacements like PressPlay, with limited catalog, high prices, and anti-copying technology (digital rights management, or DRM) that alienated users by the millions by treating them like crooks instead of customers. These half-baked ventures did untold damage to the record companies and their parent firms.

Just look at Sony: they should have been at the top of the heap. They produce some of the world's finest, best-designed electronics. They own one of the largest record labels in the world. The synergy should have been incredible. Electronics would design the Walkmen, music would take care of catalog, and marketing would sell it all.

You know the joke about European hell? The English do the cooking, the Germans are the lovers, the Italians are the police, and the French run the government. With Sony, it seemed like music was designing the Walkmen, marketing was doing the catalog, and electronics was in charge of selling. Sony's portable players — the MusicClip and others — were so crippled by anti-copying technology that they couldn't even play MP3s, and the music selection at Sony services like PressPlay was anemic, expensive, and equally hobbled. Sony isn't even a name in the portable audio market anymore — today's Walkman is an iPod.

Of course, Sony still has a record label — for now. But sales are falling, and the company is reeling from the 2005 "rootkit" debacle, where it deliberately infected eight million music CDs with a hacker tool called a rootkit, compromising over 500,000 U.S. computer networks, including military and government networks, all in a (failed) bid to stop copying of its CDs.

The public wasn't willing to wait for Sony and the rest to wake up and offer a service that was as compelling, exciting, and versatile as Napster. Instead, they flocked to a new generation of services like Kazaa and the various Gnutella networks. Kazaa's business-model was to set up offshore, on the tiny Polynesian island of Vanuatu, and bundle spyware with its software, making its profits off of fees from spyware crooks. Kazaa didn't want to pay billions for record industry licenses — they used the international legal and finance system to hopelessly snarl the RIAA's members through half a decade of wild profitability. The company was eventually brought to ground, but the founders walked away and started Skype and then Joost.

Meantime, dozens of other services had sprung up to fill Kazaa's niche — AllofMP3, the notorious Russian site, was eventually killed through intervention of the U.S. Trade Representative and the WTO, and was reborn practically the next day under a new name.

It's been eight years since Sean Fanning created Napster in his

college dorm-room. Eight years later, there isn't a single authorized music service that can compete with the original Napster.
Record sales are down every year, and digital music sales aren't
filling in the crater. The record industry has contracted to four
companies, and it may soon be three if EMI can get regulatory
permission to put itself on the block.

The sue-'em-all-and-let-God-sort-'em-out plan was a flop in
the box office, a flop in home video, and a flop overseas. So why is
Hollywood shooting a remake?

YouTube, 2007, bears some passing similarity to Napster, 2001.
Founded by a couple guys in a garage, rocketed to popular success,
heavily capitalized by a deep-pocketed giant. Its business model?
Turn popularity into dollars and offer a share to the rightsholders
whose works they're using. This is a historically sound plan:
cable operators got rich by retransmitting broadcasts without
permission, and once they were commercial successes, they sat
down to negotiate to pay for those copyrights (just as the record
companies negotiated with composers after they'd gotten rich
selling records bearing those compositions).

YouTube '07 has another similarity to Napster '01: it is being
sued by entertainment companies.

Only this time, it's not (just) the record industry. Broadcasters, movie studios, people who make video or audio are getting
in on the act. I recently met an NBC employee who told me that
he thought that a severe, punishing legal judgment would send
a message to the tech industry not to field this kind of service
anymore.

Let's hope he's wrong. Google — YouTube's owners — is a
grown-up of a company, unusual in a tech industry populated by
corporate adolescents. They have lots of money and a sober interest in keeping it. They want to sit down with A/V rightsholders

and do a deal. Six years after the Napster verdict, that kind of willingness is in short supply.

Most of the tech "companies" with an interest in commercializing Internet AV have no interest in sitting down with the studios. They're either nebulous open source projects (like mythtv, a free hyper-TiVo that skips commercials, downloads and shares videos, and is wide open to anyone who wants to modify and improve it), politically motivated anarchists (like ThePirateBay, a Swedish Bit-Torrent tracker site that has mirrors in three countries with non-interoperable legal systems, where they respond to legal notices by writing sarcastic and profane letters and putting them online), or out-and-out crooks like the bootleggers who use P2P to seed their DVD counterfeiting operations.

It's not just YouTube. TiVo, who pioneered the personal video recorder, is feeling the squeeze, being systematically locked out of the digital cable and satellite market. Their efforts to add a managed TiVoToGo service were attacked by the rightsholders who fought at the FCC to block them. Cable/satellite operators and the studios would much prefer the public to transition to "bundled" PVRs that come with your TV service.

These boxes are owned by the cable/satellite companies, who have absolute control over them. Time-Warner has been known to remotely delete stored episodes of shows just before the DVD ships, and many operators have started using "flags" that tell recorders not to allow fast-forwarding, or to prevent recording altogether.

The reason that YouTube and TiVo are more popular than ThePirateBay and mythtv is that they're the easiest way for the public to get what it wants — the video we want, the way we want it. We use these services because they're like the original Napster: easy, well-designed, functional.

But if the entertainment industry squeezes these players out,

ThePirateBay and mythtv are right there, waiting to welcome us in with open arms. ThePirateBay has already announced that it is launching a YouTube competitor with no-plugin, in-browser viewing. Plenty of entrepreneurs are looking at easing the pain and cast of setting up your own mythtv box. The only reason that the barriers to BitTorrent and mythtv exist is that it hasn't been worth anyone's while to capitalize projects to bring them down. But once the legit competitors of these services are killed, look out.

The thing is, the public doesn't want managed services with limited rights. We don't want to be stuck using approved devices in approved ways. We never have — we are the spiritual descendants of the customers for "illegal" record albums and "illegal" cable TV. The demand signal won't go away.

There's no good excuse for going into production on a sequel to The Napster Wars. We saw that movie. We know how it turns out. Every Christmas, we get articles about how this was the worst Christmas ever for CDs. You know what? CD sales are *never* going to improve. CDs have been rendered obsolete by Internet distribution — and the record industry has locked itself out of the only profitable, popular music distribution systems yet invented.

Companies like Google/YouTube and TiVo are rarities: tech companies that want to do deals. They need to be cherished by entertainment companies, not sued.

(Thanks to Bruce Nash and The-Numbers.com for research assistance with this article.)

# You DO Like Reading Off
# a Computer Screen

(Originally published in *Locus,* March 2007.)

"I don't like reading off a computer screen" — it's a cliché of the ebook world. It means "I don't read novels off of computer screens" (or phones, or PDAs, or dedicated ebook readers), and often as not the person who says it is someone who, in fact, spends every hour that Cthulhu sends reading off a computer screen. It's like watching someone shovel Mars bars into his gob while telling you how much he hates chocolate.

But I know what you mean. You don't like reading long-form works off of a computer screen. I understand perfectly — in the ten minutes since I typed the first word in the paragraph above, I've checked my mail, deleted two spams, checked an image-sharing community I like, downloaded a YouTube clip of Stephen Colbert complaining about the iPhone (pausing my MP3 player first), cleared out my RSS reader, and then returned to write this paragraph.

This is not an ideal environment in which to concentrate on long-form narrative (sorry, one sec, gotta blog this guy who's made cardboard furniture) (wait, the Colbert clip's done, gotta start the music up) (19 more RSS items). But that's not to say that it's not an *entertainment medium* — indeed, practically everything I do on the computer entertains the hell out of me. It's nearly all text-based, too. Basically, what I do on the computer is pleasure-reading. But it's a fundamentally more scattered, splintered kind of pleasure. Computers have their own cognitive style, and it's not much like the cognitive style invented with the first modern novel (one sec, let me google that and confirm it), *Don Quixote*, some 400 years ago.

The novel is an invention, one that was engendered by technological changes in information display, reproduction, and distribution. The cognitive style of the novel is different from the cognitive style of the legend. The cognitive style of the computer is different from the cognitive style of the novel.

Computers want you to do lots of things with them. Networked computers doubly so — they (another RSS item) have a million ways of asking for your attention, and just as many ways of rewarding it.

There's a persistent fantasy/nightmare in the publishing world of the advent of very sharp, very portable computer screens. In the fantasy version, this creates an infinite new market for electronic books, and we all get to sell the rights to our work all over again. In the nightmare version, this leads to runaway piracy, and no one ever gets to sell a novel again.

I think they're both wrong. The infinitely divisible copyright ignores the "decision cost" borne by users who have to decide, over and over again, whether they want to spend a millionth of a cent on a millionth of a word — no one buys newspapers by the paragraph, even though most of us only read a slim fraction of any given paper. A super-sharp, super-portable screen would be used to read all day long, but most of us won't spend most of our time reading anything recognizable as a book on them.

Take the record album. Everything about it is technologically pre-determined. The technology of the LP demanded artwork to differentiate one package from the next. The length was set by the groove density of the pressing plants and playback apparatus. The dynamic range likewise. These factors gave us the idea of the

40-to-60-minute package, split into two acts, with accompanying artwork. Musicians were encouraged to create works that would be enjoyed as a unitary whole for a protracted period — think of *Dark Side of the Moon,* or *Sgt. Pepper's.*

No one thinks about albums today. Music is now divisible to the single, as represented by an individual MP3, and then subdivisible into snippets like ringtones and samples. When recording artists demand that their works be considered as a whole — like when Radiohead insisted that the iTunes Music Store sell their whole album as a single, indivisible file that you would have to listen to all the way through — they sound like cranky throwbacks.

The idea of a 60-minute album is as weird in the Internet era as the idea of sitting through 15 hours of *Der Ring des Nibelungen* was 20 years ago. There are some anachronisms who love their long-form opera, but the real action is in the more fluid stuff that can slither around on hot wax — and now the superfluid droplets of MP3s and samples. Opera survives, but it is a tiny sliver of a much bigger, looser music market. The future composts the past: old operas get mounted for living anachronisms; Andrew Lloyd Webber picks up the rest of the business.

Or look at digital video. We're watching more digital video, sooner, than anyone imagined. But we're watching it in three-minute chunks from YouTube. The video's got a pause button so you can stop it when the phone rings and a scrubber to go back and forth when you miss something while answering an IM.

And attention spans don't increase when you move from the PC to a handheld device. These things have less capacity for multitasking than real PCs, and the network connections are slower and more expensive. But they are fundamentally multitasking devices — you can always stop reading an ebook to play a hand of solitaire that is interrupted by a phone call — and their social context is that they are used in public places, with a million distractions. It is socially acceptable to interrupt someone who is looking at a PDA screen. By contrast, the TV room — a whole room for TV! — is a shrine where none may speak until the commercial airs.

The problem, then, isn't that screens aren't sharp enough to read novels off of. The problem is that novels aren't screeny enough to warrant protracted, regular reading on screens.

Electronic books are a wonderful adjunct to print books. It's great to have a couple hundred novels in your pocket when the plane doesn't take off or the line is too long at the post office. It's cool to be able to search the text of a novel to find a beloved passage. It's excellent to use a novel socially, sending it to your friends, pasting it into your sig file.

But the numbers tell their own story — people who read off of screens all day long buy lots of print books and read them primarily on paper. There are some who prefer an all-electronic existence (I'd like to be able to get rid of the objects after my first reading, but keep the ebooks around for reference), but they're in a tiny minority.

There's a generation of web-writers who produce "pleasure reading" on the Web. Some are funny. Some are touching. Some are enraging. Most dwell in Sturgeon's 90th percentile and below. They're not writing novels. If they were, they wouldn't be Web-writers.

Mostly, we can read just enough of a free ebook to decide whether to buy it in hardcopy — but not enough to substitute the ebook for the hardcopy. Like practically everything in marketing and promotion, the trick is to find the form of the work that serves as enticement, not replacement.

Sorry, got to go — 8 more emails.

# How Do You Protect Artists?

(Originally published as "Online Censorship Hurts Us All," *The Guardian*, October 2, 2007.)

Artists have lots of problems. We get plagiarized, ripped off by publishers, savaged by critics, counterfeited — and we even get our works copied by "pirates" who give our stuff away for free online.

But no matter how bad these problems get, they're a distant second to the gravest, most terrifying problem an artist can face: censorship.

It's one thing to be denied your credit or compensation, but it's another thing entirely to have your work suppressed, burned, or banned. You'd never know it, however, judging from the state of the law surrounding the creation and use of Internet publishing tools.

Since 1995, every single legislative initiative on this subject in the UK's parliament, the European parliament, and the U.S. Congress has focused on making it easier to suppress "illegitimate" material online. From libel to copyright infringement, from child porn to anti-terror laws, our legislators have approached the Internet with a single-minded focus on seeing to it that bad material is expeditiously removed.

And that's the rub. I'm certainly no fan of child porn or hate speech, but every time a law is passed that reduces the burden of proof on those who would remove material from the Internet, artists' fortunes everywhere are endangered.

Take the U.S.'s 1998 Digital Millennium Copyright Act, which has equivalents in every European state that has implemented the 2001 European Union Copyright Directive. The DMCA allows anyone to have any document on the Internet removed, simply by contacting its publisher and asserting that the work infringes his copyright.

The potential for abuse is obvious, and the abuse has been widespread: from the Church of Scientology to companies that don't like what reporters write about them, DMCA takedown notices have fast become the favorite weapon in the cowardly bully's arsenal.

But takedown notices are just the start. While they can help silence critics and suppress timely information, they're not actually very effective at stopping widespread copyright infringement. Viacom sent over 100,000 takedown notices to YouTube last February, but seconds after it was all removed, new users uploaded it again.

Even these takedown notices were sloppily constructed: they included videos of friends eating at barbecue restaurants and videos of independent bands performing their own work. As a Recording Industry Association of America spokesman quipped, "When you go trawling with a net, you catch a few dolphins."

Viacom and others want hosting companies and online service providers to preemptively evaluate all the material that their users put online, holding it to ensure that it doesn't infringe copyright before they release it.

This notion is impractical in the extreme, for at least two reasons. First, an exhaustive list of copyrighted works would be unimaginably huge, as every single creative work is copyrighted from the instant that it is created and "fixed in a tangible medium."

Second, even if such a list did exist, it would be trivial to defeat, simply by introducing small changes to the infringing copies, as spammers do with the text of their messages in order to evade spam filters.

In fact, the spam wars have some important lessons to teach us here. Like copyrighted works, spams are infinitely varied and more are being created every second. Any company that could identify spam messages — including permutations and variations on existing spams — could write its own ticket to untold billions.

Some of the smartest, most dedicated engineers on the planet devote every waking hour to figuring out how to spot spam before it gets delivered. If your inbox is anything like mine, you'll agree that the war is far from won.

If the YouTubes of the world are going to prevent infringement, they're going to have to accomplish this by hand-inspecting every one of the tens of billions of blog posts, videos, text files, music files, and software uploads made to every single server on the Internet.

And not just cursory inspections, either — these inspections will have to be undertaken by skilled, trained specialists (who'd better be talented linguists, too — how many English speakers can spot an infringement in Urdu?).

Such experts don't come cheap, which means that you can anticipate a terrible denuding of the fertile jungle of Internet hosting companies that are primary means by which tens of millions of creative people share the fruits of their labor with their fans and colleagues.

It would be a great Sovietization of the world's digital printing presses, a contraction of a glorious anarchy of expression into a regimented world of expensive and narrow venues for art.

It would be a death knell for the kind of focused, noncommercial material whose authors couldn't fit the bill for a "managed" service's legion of lawyers, who would be replaced by more of the same — the kind of lowest common denominator rubbish that fills the cable channels today.

And the worst of it is, we're marching toward this "solution" in the name of protecting artists. Gee, thanks.

# It's the Information Economy, Stupid

(Originally published as "Free Data Sharing Is Here to Stay," *The Guardian*, September 18, 2007.)

Since the 1970s, pundits have predicted a transition to an "information economy." The vision of an economy based on information seized the imaginations of the world's governments. For decades now, they have been creating policies to "protect" information — stronger copyright laws, international treaties on patents and trademarks, treaties to protect anti-copying technology.

The thinking is simple: an information economy must be based on buying and selling information. Therefore, we need policies to make it harder to get access to information unless you've paid for it. That means that we have to make it harder for you to share information, even after you've paid for it. Without the ability to fence off your information property, you can't have an information market to fuel the information economy.

But this is a tragic case of misunderstanding a metaphor. Just as the industrial economy wasn't based on making it harder to get access to machines, the information economy won't be based on making it harder to get access to information. Indeed, the opposite seems to be true: the more IT we have, the easier it is to access any given piece of information — for better or for worse.

It used to be that copy-prevention companies' strategies went like this: "We'll make it easier to buy a copy of this data than to make an unauthorized copy of it. That way, only the *über*-nerds and the cash-poor/time-rich classes will bother to copy instead of buy." But every time a PC is connected to the Internet and its owner is taught to use search tools like Google (or The Pirate Bay), a third option appears: you can just download a copy from the Internet. Every techno-literate participant in the information economy can choose to access any data, without having to break the anti-copying technology, just by searching for the cracked copy on the public Internet. If there's one thing we can be sure of, it's that an information economy will increase the technological literacy of its participants.

As I write this, I am sitting in a hotel room in Shanghai, behind the Great Firewall of China. Theoretically, I can't access blogging services that carry negative accounts of Beijing's doings, like WordPress, Blogger, and LiveJournal, nor the image-sharing site Flickr, nor Wikipedia. The (theoretically) omnipotent bureaucrats of the local Minitrue have deployed their finest engineering talent to stop me. Well, these cats may be able to order political prisoners executed and their organs harvested for Party members, but they've totally failed to keep Chinese people (and big-nose tourists like me) off the world's Internet. The WTO is rattling its sabers at China today, demanding that they figure out how to stop Chinese people from looking at Bruce Willis movies without permission — but the Chinese government can't even figure out how to stop Chinese people from looking at seditious revolutionary tracts online.

And, of course, as Paris Hilton, the Church of Scientology, and the King of Thailand have discovered, taking a piece of information off the Internet is like getting food coloring out of a swimming pool. Good luck with that.

To see the evidence of the real information economy, look to all the economic activity that the Internet enables — not the stuff that it impedes. All the commerce conducted by salarymen who can book their own flights with Expedia instead of playing blind-man's bluff with a travel agent. ("Got any flights after 4 PM to Frankfurt?") All the garage crafters selling their goods on Etsy.com. All the publishers selling obscure books through Amazon that no physical bookstore was willing to carry. The *salwar kameez* tailors in India selling bespoke clothes to westerners via eBay, without intervention by a series of skimming intermediaries. The Internet-era musicians who use the Net to pack venues all over the world by giving away their recordings on social services like MySpace. Hell, look at my last barber, in Los Angeles: the man doesn't use a PC, but I found him by googling for "barbers" with my postcode — the information economy is driving his cost of customer acquisition to zero, and he doesn't even have to actively participate in it.

Better access to more information is the hallmark of the information economy. The more IT we have, the more skill we have, the faster our networks get, and the better our search tools get, the more economic activity the information economy generates. Many of us sell information in the information economy — I sell my printed books by giving away electronic books, lawyers and architects and consultants are in the information business and they drum up trade with Google ads, and Google is nothing but an info-broker — but none of us rely on curtailing access to information. Like a bottled water company, we compete with free by supplying a superior service, not by eliminating the competition.

The world's governments might have bought into the old myth of the information economy, but not so much that they're willing to ban the PC and the Internet.

# Downloads Give Amazon Jungle Fever

(Originally published in *The Guardian*, December 11, 2007.)

Let me start by saying that I love Amazon. I buy everything from books to clothes to electronics to medication to food to batteries to toys to furniture to baby supplies from the company. I once even bought an ironing board on Amazon. No company can top them for ease of use or for respecting consumer rights when it comes to refunds, ensuring satisfaction, and taking good care of loyal customers.

As a novelist, I couldn't be happier about Amazon's existence. Not only does Amazon have a set of superb recommendation tools that help me sell books, but it also has an affiliate program that lets me get up to 8.5 percent in commissions for sales of my books through the site — nearly doubling my royalty rate.

As a consumer advocate and activist, I'm delighted by almost every public policy initiative from Amazon. When the Author's Guild tried to get Amazon to curtail its used-book market, the company refused to back down. Founder Jeff Bezos (who is a friend of mine) even wrote, "When someone buys a book, they are also buying the right to resell that book, to loan it out, or to even give it away if they want. Everyone understands this."

More recently, Amazon stood up to the U.S. government, who'd gone on an illegal fishing expedition for terrorists (TERRORISTS! TERRORISTS! TERRORISTS!) and asked Amazon to turn over the purchasing history of 24,000 Amazon customers. The company spent a fortune fighting for our rights, and won.

It also has a well-deserved reputation for taking care over

copyright "takedown" notices for the material that its customers post on its site, discarding ridiculous claims rather than blindly acting on every single notice, no matter how frivolous.

But for all that, it has to be said: Whenever Amazon tries to sell a digital download, it turns into one of the dumbest companies on the Web.

Take the Kindle, the $400 handheld ebook reader that Amazon shipped recently, to vast, ringing indifference.

The device is cute enough — in a clumsy, overpriced, generation-one kind of way — but the early adopter community recoiled in horror at the terms of service and anti-copying technology that infected it. Ebooks that you buy through the Kindle can't be lent or resold (remember, "When someone buys a book, they are also buying the right to resell that book... Everyone understands this.")

Mark Pilgrim's "The Future of Reading" enumerates five other Kindle showstoppers: Amazon can change your ebooks without notifying you or getting your permission; and if you violate any of the "agreement," it can delete your ebooks, even if you've paid for them, and you get no appeal.

It's not just the Kindle, either. Amazon Unbox, the semi-abortive video download service, shipped with terms of service that included your granting permission for Amazon to install any software on your computer, to spy on you, to delete your videos, to delete any other file on your hard drive, to deny you access to your movies if you lose them in a crash. This comes from the company that will cheerfully ship you a replacement DVD if you email them and tell them that the one you just bought never turned up in the post.

Even Amazon's much-vaunted MP3 store comes with terms of service that prevent lending and reselling.

I am mystified by this. Amazon is the kind of company that

every etailer should study and copy — the gold standard for e-commerce. You'd think that if there was any company that would intuitively get the Web, it would be Amazon.

What's more, this is a company that stands up to rightsholder groups, publishers, and the U.S. government — but only when it comes to physical goods. Why is it that whenever a digital sale is in the offing, Amazon rolls over on its back and wets itself?

# What's the Most Important Right Creators Have?

(Originally published as "How Big Media's Copyright Campaigns Threaten Internet Free Expression," *InformationWeek*, November 5, 2007.)

Any discussion of "creator's rights" is likely to be limited to talk about copyright, but copyright is just a side-dish for creators: the most important right we have is the right to free expression. And these two rights are always in tension.

Take Viacom's claims against YouTube. The entertainment giant says that YouTube has been profiting from the fact that YouTube users upload clips from Viacom shows, and they demand that YouTube take steps to prevent this from happening in the future. YouTube actually offered to do something very like this: they invited Viacom and other rightsholders to send them all the clips they wanted kept offline, and promised to programmatically detect these clips and interdict them.

But Viacom rejected this offer. Rather, the company wants YouTube to just figure it out, determine a priori which video clips are being presented with permission and which ones are not. After all, Viacom does the very same thing: it won't air clips until a battalion of lawyers have investigated them and determined whether they are lawful.

But the Internet is not cable television. Net-based hosting outfits—including YouTube, Flickr, Blogger, Scribd, and the Internet Archive — offer free publication venues to all comers, enabling anyone to publish anything. In 1998's Digital Millennium Copyright Act, Congress considered the question of liability for these companies and decided to offer them a mixed deal: hosting companies don't need to hire a million lawyers to review every blog-post before it goes live, but rightsholders can order them to remove any infringing material from the Net just by sending them a notice that the material infringes.

This deal enabled hosting companies to offer free platforms for publication and expression to everyone. But it also allowed anyone to censor the Internet, just by making claims of infringement, without offering any evidence to support those claims, without having to go to court to prove their claims (this has proven to be an attractive nuisance, presenting an irresistible lure to anyone with a beef against an online critic, from the Church of Scientology to Diebold's voting machines division).

The proposal for online hosts to figure out what infringes and what doesn't is wildly impractical. Under most countries' copyright laws, creative works receive a copyright from the moment that they are "fixed in a tangible medium" (hard drives count), and this means that the pool of copyrighted works is so large as to be, practically speaking, infinite. Knowing whether a work is copyrighted, who holds the copyright, and whether a posting is made with the rightsholder's permission (or in accord with each nation's varying ideas about fair use) is impossible. The only way to be sure is to start from the presumption that each creative work is infringing, and then make each Internet user prove, to some lawyer's satisfaction, that she has the right to post each drib of content that appears on the Web.

Imagine that such a system were the law of the land. There's no way Blogger or YouTube or Flickr could afford to offer free hosting to their users. Rather, all these hosted services would have to charge enough for access to cover the scorching legal bills associated with checking all material. And not just the freebies, either: your local ISP, the servers hosting your company's website, or your page for family genealogy: they'd all have to do the same kind of continuous checking and re-checking of every file you publish with them.

It would be the end of any publication that couldn't foot the legal bills to get off the ground. The multi-billion-page Internet would collapse into the homogeneous world of cable TV (remember when we thought that a "500-channel universe" would be unimaginably broad? Imagine an Internet with only 500 "channels"!). From Amazon to Ask A Ninja, from Blogger to the Everlasting Blort, every bit of online content is made possible by removing the cost of paying lawyers to act as the Internet's gatekeepers.

This is great news for artists. The traditional artist's lament is that our publishers have us over a barrel, controlling the narrow and vital channels for making works available — from big gallery owners to movie studios to record labels to New York publishers. That's why artists have such a hard time negotiating a decent deal for themselves (for example, most beginning recording artists have to agree to have money deducted from their royalty statements for "breakage" of records en route to stores — and these deductions are also levied against digital sales through the iTunes Store!).

But, thanks to the Web, artists have more options than ever. The Internet's most popular video podcasts aren't associated with TV networks (with all the terrible, one-sided deals that would entail), rather, they're independent programs like RocketBoom, Homestar Runner, or the late, lamented Ze Frank Show. These creators — along with all the musicians, writers, and other artists using the Net to earn their living — were able to write their own ticket. Today, major artists like Radiohead and Madonna are leaving the record labels behind and trying novel, Net-based ways of promoting their work.

And it's not just the indies who benefit: the existence of successful independent artists creates fantastic leverage for artists who negotiate with the majors. More and more, the big media companies' "like it or leave it" bargaining stance is being undermined by the possibility that the next big star will shrug, turn on her heel, and make her fortune without the big companies' help. This has humbled the bigs, making their deals better and more artist-friendly.

Bargaining leverage is just for starters. The greatest threat that art faces is suppression. Historically, artists have struggled just to make themselves heard, just to safeguard the right to express themselves. Censorship is history's greatest enemy of art. A limited-liability Web is a Web where anyone can post anything and reach *everyone*.

What's more, this privilege isn't limited to artists. All manner of communication, from the personal introspection in public "diaries" to social chatter on MySpace and Facebook, are now possible. Some artists have taken the bizarre stance that this "trivial" matter is unimportant and thus a poor excuse for allowing hosted services to exist in the first place. This is pretty arrogant: a society where only artists are allowed to impart "important" messages and where the rest of us are supposed to shut up about our loves, hopes, aspirations, jokes, family, and wants is hardly a democratic paradise.

Artists are in the free expression business, and technology that helps free expression helps artists. When lowering the cost of copyright enforcement raises the cost of free speech, every artist has a duty to speak out. Our ability to make our art is inextricably linked with the billions of Internet users who use the network to talk about their lives.

# Giving it Away

(Originally published in *Forbes*, December 2006.)

I've been giving away my books ever since my first novel came out, and boy has it ever made me a bunch of money.

When my first novel, *Down and Out in the Magic Kingdom,* was published by Tor Books in January 2003, I also put the entire electronic text of the novel on the Internet under a Creative Commons license that encouraged my readers to copy it far and wide. Within a day, there were 30,000 downloads from my site (and those downloaders were in turn free to make more copies). Three years and six printings later, more than 700,000 copies of the book have been downloaded from my site. The book's been translated into more languages than I can keep track of, key concepts from it have been adopted for software projects, and there are two competing fan audio adaptations online.

Most people who download the book don't end up buying it, but they wouldn't have bought it in any event, so I haven't lost any sales, I've just won an audience. A tiny minority of downloaders treat the free ebook as a substitute for the printed book — those are the lost sales. But a much larger minority treat the ebook as an enticement to buy the printed book. They're gained sales. As long as gained sales outnumber lost sales, I'm ahead of the game. After all, distributing nearly a million copies of my book has cost me nothing.

The thing about an ebook is that it's a social object. It wants to be copied from friend to friend, beamed from a Palm device, pasted into a mailing list. It begs to be converted to witty signatures at the bottom of emails. It is so fluid and intangible that it can spread itself over your whole life. Nothing sells books like a personal recommendation — when I worked in a bookstore, the sweetest words we could hear were "My friend suggested I pick up...." The friend had made the sale for us, we just had to consummate it. In an age of online friendship, ebooks trump dead trees for word of mouth.

There are two things that writers ask me about this arrangement: First, does it sell more books, and second, how did you talk your publisher into going for this mad scheme?

There's no empirical way to prove that giving away books sells more books — but I've done this with three novels and a short story collection (and I'll be doing it with two more novels and another collection in the next year), and my books have consistently outperformed my publisher's expectations. Comparing their sales to the numbers provided by colleagues suggests that they perform somewhat better than other books from similar writers at similar stages in their careers. But short of going back in time and re-releasing the same books under the same circumstances without the free ebook program, there's no way to be sure.

What is certain is that every writer who's tried giving away ebooks to sell books has come away satisfied and ready to do it some more.

How did I talk Tor Books into letting me do this? It's not as if Tor is a spunky dotcom upstart. They're the largest science fiction publisher in the world, and they're a division of the German publishing giant Holtzbrinck. They're not patchouli-scented info-hippies who believe that information wants to be free. Rather, they're canny assessors of the world of science fiction, perhaps the most social of all literary genres. Science fiction is driven by organized fandom, volunteers who put on hundreds of literary conventions in every corner of the globe, every weekend of the

year. These intrepid promoters treat books as markers of identity and as cultural artifacts of great import. They evangelize the
books they love, form subcultures around them, cite them in political arguments, sometimes they even rearrange their lives and
jobs around them.

What's more, science fiction's early adopters defined the social
character of the Internet itself. Given the high correlation between technical employment and science fiction reading, it was
inevitable that the first nontechnical discussion on the Internet
would be about science fiction. The online norms of idle chatter,
fannish organizing, publishing, and leisure are descended from sf
fandom, and if any literature has a natural home in cyberspace,
it's science fiction, the literature that coined the very word "cyberspace."

Indeed, science fiction was the first form of widely pirated
literature online, through "bookwarez" channels that contained
books that had been hand-scanned, a page at a time, converted to
digital text, and proof-read. Even today, the most widely pirated
literature online is sf.

Nothing could make me more sanguine about the future. As
publisher Tim O'Reilly wrote in his seminal essay, "Piracy is Progressive Taxation," "being well-enough known to be pirated [is] a
crowning achievement." I'd rather stake my future on a literature
that people care about enough to steal than devote my life to a
form that has no home in the dominant medium of the century.

What about that future? Many writers fear that in the future,
electronic books will come to substitute more readily for print
books, due to changing audiences and improved technology. I am
skeptical of this — the codex format has endured for centuries
as a simple and elegant answer to the affordances demanded by
print, albeit for a relatively small fraction of the population. Most
people aren't and will never be readers — but the people who are

readers will be readers forever, and they are positively pervy for paper.

But say it does come to pass that electronic books are all anyone wants.

I don't think it's practical to charge for copies of electronic works. Bits aren't ever going to get harder to copy. So we'll have to figure out how to charge for something else. That's not to say you can't charge for a copy-able bit, but you sure can't force a reader to pay for access to information anymore.

This isn't the first time creative entrepreneurs have gone through one of these transitions. Vaudeville performers had to transition to radio, an abrupt shift from having perfect control over who could hear a performance (if they don't buy a ticket, you throw them out) to no control whatsoever (any family whose twelve-year-old could build a crystal set, the day's equivalent of installing file-sharing software, could tune in). There were business-models for radio, but predicting them a priori wasn't easy. Who could have foreseen that radio's great fortunes would be had through creating a blanket license, securing a Congressional consent decree, chartering a collecting society, and inventing a new form of statistical mathematics to fund it?

Predicting the future of publishing — should the wind change and printed books become obsolete — is just as hard. I don't know how writers would earn their living in such a world, but I *do* know that I'll never find out by turning my back on the Internet. By being in the middle of electronic publishing, by watching what hundreds of thousands of my readers do with my ebooks, I get better market intelligence than I could through any other means. As does my publisher. As serious as I am about continuing to work as a writer for the foreseeable future, Tor Books and Holtzbrinck are just as serious. They've got even more riding on the future of publishing than I do. So when I approached my publisher with this plan to give away books to sell books, it was a no-brainer for them.

It's good business for me, too. This "market research" of giving away ebooks sells printed books. What's more, having my books more widely read opens many other opportunities for me to earn a living from activities around my writing, such as the Fulbright Chair I got at USC this year, this high-paying article in *Forbes*, speaking engagements and other opportunities to teach, write and license my work for translation and adaptation. My fans' tireless evangelism for my work doesn't just sell books — it sells *me*.

The golden age of hundreds of writers who lived off of nothing but their royalties is bunkum. Throughout history, writers have relied on day jobs, teaching, grants, inheritances, translation, licensing, and other varied sources to make ends meet. The Internet not only sells more books for me, it also gives me more opportunities to earn my keep through writing-related activities.

There has never been a time when more people were reading more words by more authors. The Internet is a literary world of written words. What a fine thing that is for writers.

# Science Fiction Is the Only Literature People Care Enough About to Steal on the Internet

(Originally published in *Locus*, July 2006.)

As a science fiction writer, no piece of news could make me more hopeful. It beats the hell out of the alternative — a future where the dominant, pluripotent, ubiquitous medium has no place for science fiction literature.

When radio and records were invented, they were pretty bad news for the performers of the day. Live performance demanded charisma, the ability to really put on a magnetic show in front of a crowd. It didn't matter how technically accomplished you were: if you stood like a statue on stage, no one wanted to see you do your thing. On the other hand, you succeeded as a mediocre player, provided you attacked your performance with a lot of brio.

Radio was clearly good news for musicians — lots more musicians were able to make lots more music, reaching lots more people and making lots more money. It turned performance into an industry, which is what happens when you add technology to art. But it was terrible news for charismatics. It put them out on the street, stuck them with flipping burgers and driving taxis. They knew it, too. Performers lobbied to have the Marconi radio banned, to send Marconi back to the drawing board, charged with inventing a radio they could charge admission to. "We're charismatics, we do something as old and holy as the first story told before the first fire in the first cave. What right have you to insist that we should become mere clerks, working in an obscure back-room, leaving you to commune with our audiences on our behalf?"

Technology giveth and technology taketh away. Seventy years later, Napster showed us that, as William Gibson noted, "We may be at the end of the brief period during which it is possible to charge for recorded music." Surely we're at the end of the period where it's possible to exclude those who don't wish to pay. Every song released can be downloaded gratis from a peer-to-peer network (and will shortly get easier to download, as hard drive price-performance curves take us to a place where all the music ever recorded will fit on a disposable pocket-drive that you can just walk over to a friend's place and copy).

But have no fear: the Internet makes it possible for recording artists to reach a wider audience than ever dreamt of before. Your potential fans may be spread in a thin, even coat over the world, in a configuration that could never be cost-effective to reach with traditional marketing. But the Internet's ability to lower the costs for artists to reach their audiences and for audiences to find artists suddenly renders possible more variety in music than ever before.

Those artists can use the Internet to bring people back to the live performances that characterized the heyday of Vaudeville. Use your recordings — which you can't control — to drive admissions to your performances, which you can control. It's a model that's worked great for jam bands like the Grateful Dead and Phish. It's also a model that won't work for many of today's artists; seventy years of evolutionary pressure has selected for artists who are more virtuoso than charismatic, artists optimized for recording-based income instead of performance-based income. "How dare you tell us that we are to be trained monkeys, capering on a stage for your amusement? We're not charismatics, we're white-collar workers. We commune with our muses behind closed doors and deliver up our work product when it's done, through plastic, laser-etched discs. You have no right to demand that we convert to a live-performance economy."

Technology giveth and technology taketh away. As bands on MySpace — who can fill houses and sell hundreds of thousands of discs without a record deal, by connecting individually with fans — have shown, there's a new market aborning on the Internet for music, one with fewer gatekeepers to creativity than ever before.

That's the purpose of copyright, after all: to decentralize who gets to make art. Before copyright, we had patronage: you could make art if the Pope or the king liked the sound of it. That produced some damned pretty ceilings and frescos, but it wasn't until control of art was given over to the market — by giving publishers a monopoly over the works they printed, starting with the Statute of Anne in 1709 — that we saw the explosion of creativity that investment-based art could create. Industrialists weren't great arbiters of who could and couldn't make art, but they were better than the Pope.

The Internet is enabling a further decentralization in who gets to make art, and like each of the technological shifts in cultural production, it's good for some artists and bad for others. The important question is: Will it let more people participate in cultural production? Will it further decentralize decision-making for artists?

And for sf writers and fans, the further question is: "Will it be any good to our chosen medium?" Like I said, science fiction is the only literature people care enough about to steal on the Internet. It's the only literature that regularly shows up, scanned and run through optical character recognition software and lovingly hand-edited on darknet newsgroups, Russian websites, IRC channels, and elsewhere (yes, there's also a brisk trade in comics and technical books, but I'm talking about prose fiction here — though this is clearly a sign of hope for our friends in tech publishing and funnybooks).

Some writers are using the Internet's affinity for sf to great effect. I've released every one of my novels under Creative Commons licenses that encourage fans to share them freely and widely — even, in some cases, to remix them and to make new editions of them for use in the developing world. My first novel, *Down and Out in the Magic Kingdom,* is in its sixth printing from Tor, and has been downloaded more than 650,000 times from my website, and an untold number of times from others' websites.

I've discovered what many authors have also discovered: releasing electronic texts of books drives sales of the print editions. An sf writer's biggest problem is obscurity, not piracy. Of all the people who chose not to spend their discretionary time and cash on our works today, the great bulk of them did so because they didn't know they existed, not because someone handed them a free ebook version.

But what kind of artist thrives on the Internet? Those who can establish a personal relationship with their readers — something science fiction has been doing for as long as pros have been hanging out in the con suite instead of the green room. These conversational artists come from all fields, and they combine the best aspects of charisma and virtuosity with charm — the ability to conduct their online selves as part of a friendly salon that establishes a non-substitutable relationship with their audiences. You might find a film, a game, and a book to be equally useful diversions on a slow afternoon, but if the novel's author is a pal of yours, that's the one you'll pick. It's a competitive advantage that can't be beat.

See Neil Gaiman's blog, where he manages the trick of carrying on a conversation with millions. Or Charlie Stross's Usenet posts.

Scalzi's blogs. J. Michael Straczynski's presence on Usenet — while in production on *Babylon 5*, no less — breeding an army of rabid fans ready to fax-bomb recalcitrant TV execs into submission and syndication. See also the MySpace bands selling a million units of their CDs by adding each buyer to their "friends lists." Watch Eric Flint manage the Baen Bar, and Warren Ellis's good-natured growling on his sites, lists, and so forth.

Not all artists have in them to conduct an online salon with their audiences. Not all Vaudevillians had it in them to transition to radio. Technology giveth and technology taketh away. Sf writers are supposed to be soaked in the future, ready to come to grips with it. The future is conversational: when there's more good stuff that you know about that's one click away or closer than you will ever click on, it's not enough to know that some book is good. The least substitutable good in the Internet era is the personal relationship.

Conversation, not content, is king. If you were stranded on a desert island and you opted to bring your records instead of your friends, we'd call you a sociopath. Science fiction writers who can insert themselves into their readers' conversations will be set for life.

# How Copyright Broke

(Originally published in *Locus*, September 2006.)

The theory is that if the Internet can't be controlled, then copyright is dead. The thing is, the Internet is a machine for copying things cheaply, quickly, and with as little control as possible, while copyright is the right to control who gets to make copies, so these two abstractions seem destined for a fatal collision, right?

Wrong.

The idea that copyright confers the exclusive right to control copying, performance, adaptation, and general use of a creative work is a polite fiction that has been mostly harmless throughout its brief history, but which has been laid bare by the Internet, and the disjoint is showing.

Theoretically, if I sell you a copy of one of my novels, I'm conferring upon you a property interest in a lump of atoms — the pages of the book — as well as a license to make some reasonable use of the ethereal ideas embedded upon the page, the copyrighted work.

Copyright started with a dispute between Scottish and English publishers, and the first copyright law, 1709's Statute of Anne, conferred the exclusive right to publish new editions of a book on the copyright holder. It was a fair competition statute, and it was silent on the rights that the copyright holder had in respect of his customers: the readers. Publishers got a legal tool to fight their competitors, a legal tool that made a distinction between the corpus — a physical book — and the spirit — the novel writ on its pages. But this legal nicety was not "customer-facing." As far as a reader was concerned, once she bought a book, she got the same rights to it as she got to any other physical object, like a potato or a shovel. Of course, the reader couldn't print a new edition, but this had as much to do with the realities of technology as it did with the law. Printing presses were rare and expensive: telling a seventeenth-century reader that he wasn't allowed to print a new edition of a book you sold him was about as meaningful as telling him he wasn't allowed to have it laser-etched on the surface of the moon. Publishing books wasn't something readers *did*.

Indeed, until the photocopier came along, it was practically impossible for a member of the audience to infringe copyright in a way that would rise to legal notice. Copyright was like a tank-mine, designed only to go off when a publisher or record company or radio station rolled over it. We civilians *couldn't* infringe copyright (many thanks to Jamie Boyle for this useful analogy).

It wasn't the same for commercial users of copyrighted works. For the most part, a radio station that played a record was expected to secure permission to do so (though this permission usually comes in the form of a government-sanctioned blanket license that cuts through all the expense of negotiating in favor of a single monthly payment that covers all radio play). If you shot a movie, you were expected to get permission for the music you put in it. Critically, there are many uses that commercial users *never* paid for. Most workplaces don't pay for the music their employees enjoy while they work. An ad agency that produces a demo reel of recent commercials to use as part of a creative briefing to a designer doesn't pay for this extremely commercial use. A film company whose set-designer clips and copies from magazines and movies to produce a "mood book" never secures permission nor offers compensation for these uses.

Theoretically, the contours of what you may and may not do without permission are covered under a legal doctrine called "fair use," which sets out the factors a judge can use to weigh the question of whether an infringement should be punished. While fair use is a vital part of the way that works get made and used, it's very rare for an unauthorized use to get adjudicated on this basis.

No, the realpolitik of unauthorized use is that users are not required to secure permission for uses that the rightsholder will never discover. If you put some magazine clippings in your mood book, the magazine publisher will never find out you did so. If you stick a Dilbert cartoon on your office-door, Scott Adams will never know about it.

So while technically the law has allowed rightsholders to infinitely discriminate among the offerings they want to make — Special discounts on this book, which may only be read on Wednesdays! This film half-price, if you agree only to show it to people whose names start with D! — practicality has dictated that licenses could only be offered on enforceable terms.

When it comes to retail customers for information goods — readers, listeners, watchers — this whole license abstraction falls flat. No one wants to believe that the book he's brought home is only partly his, and subject to the terms of a license set out on the flyleaf. You'd be a flaming jackass if you showed up at a con and insisted that your book may not be read aloud, nor photocopied in part and marked up for a writers' workshop, nor made the subject of a piece of fan-fiction.

At the office, you might get a sweet deal on a coffee machine on the promise that you'll use a certain brand of coffee, and even sign off on a deal to let the coffee company check in on this from time to time. But no one does this at home. We instinctively and rightly recoil from the idea that our personal, private dealings in our homes should be subject to oversight from some company from whom we've bought something. We bought it. It's ours.

Even when we rent things, like cars, we recoil from the idea that Hertz might track our movements, or stick a camera in the steering wheel.

When the Internet and the PC made it possible to sell a lot of purely digital "goods" — software, music, movies, and books delivered as pure digits over the wire, without a physical good changing hands — the copyright lawyers groped about for a way to take account of this. It's in the nature of a computer that it copies what you put on it. A computer is said to be working, and of high quality, in direct proportion to the degree to which it swiftly and accurately copies the information that it is presented with.

The copyright lawyers had a versatile hammer in their toolbox: the copyright license. These licenses had been presented to corporations for years. Frustratingly (for the lawyers), these corporate customers had their own counsel, and real bargaining power, which made it impossible to impose really *interesting* conditions on them, like limiting the use of a movie such that it couldn't be fast-forwarded, or preventing the company from letting more than one employee review a journal at a time.

Regular customers didn't have lawyers or negotiating leverage. They were a natural for licensing regimes. Have a look at the next click-through "agreement" you're provided with on purchasing a piece of software or an electronic book or song. The terms set out in those agreements are positively Dickensian in their marvelous idiocy. Sony BMG recently shipped over eight million music CDs with an "agreement" that bound its purchasers to destroy their music if they left the country or had a house-fire, and to promise not to listen to their tunes while at work.

But customers understand property — you bought it, you own it — and they don't understand copyright. Practically no one understands copyright. I know editors at multibillion-dollar publishing houses who don't know the difference between copyright and trademark (if you've ever heard someone say, "You need to defend a copyright or you lose it," you've found one of these people who confuse copyright and trademark; what's more, this statement isn't particularly true of trademark, either). I once got into an argument with a senior Disney TV exec who truly believed that if you re-broadcasted an old program, it was automatically re-copy-righted and got another ninety-five years of exclusive use (that's wrong).

So this is where copyright breaks: When copyright lawyers try to treat readers and listeners and viewers as if they were (weak and unlucky) corporations who could be strong-armed into li-cense agreements you wouldn't wish on a dog. There's no con-ceivable world in which people are going to tiptoe around the property they've bought and paid for, re-checking their licenses to make sure that they're abiding by the terms of an agreement they doubtless never read. Why read something if it's non-nego-tiable, anyway?

The answer is simple: treat your readers' property as property. What readers do with their own equipment, as private, noncom-mercial actors, is not a fit subject for copyright regulation or over-sight. The Securities Exchange Commission doesn't impose rules on you when you loan a friend five bucks for lunch. Anti-gambling laws aren't triggered when you bet your kids an ice-cream cone that you'll bicycle home before them. Copyright shouldn't come between an end-user of a creative work and her property.

Of course, this approach is made even simpler by the fact that practically every customer for copyrighted works *already op-erates* on this assumption. Which is not to say that this might make some business models more difficult to pursue. Obviously, if there was some way to ensure that a given publisher was the only source for a copyrighted work, that publisher could hike up its prices, devote less money to service, and still sell its wares.

Having to compete with free copies handed from user to user makes life harder — hasn't it always?

But it is most assuredly possible. Look at Apple's wildly popular iTunes Music Store, which has sold over one billion tracks since 2003. Every song on iTunes is available as a free download from user-to-user, peer-to-peer networks like Kazaa. Indeed, the P2P monitoring company Big Champagne reports that the average time-lapse between an iTunes-exclusive song being offered by Apple and that same song being offered on P2P networks is *180 seconds*.

Every iTunes customer could readily acquire every iTunes song for free, using the fastest-adopted technology in history. Many of them do (just as many fans photocopy their favorite stories from magazines and pass them around to friends). But Apple has figured out how to compete well enough by offering a better service and a better experience to realize a good business out of this. (Apple also imposes ridiculous licensing restrictions, but that's a subject for a future column.)

Science fiction is a genre of clear-eyed speculation about the future. It should have no place for wishful thinking about a world where readers willingly put up with the indignity of being treated as "licensees" instead of customers.

# In Praise of Fanfic

(Originally published in *Locus*, May 2007.)

I wrote my first story when I was six. It was 1977, and I had just had my mind blown clean out of my skull by a new movie called *Star Wars* (the golden age of science fiction is twelve; the golden age of cinematic science fiction is six). I rushed home and stapled a bunch of paper together, trimmed the sides down so that it approximated the size and shape of a mass-market paperback, and set to work. I wrote an elaborate, incoherent ramble about *Star Wars*, in which the events of the film replayed themselves, tweaked to suit my tastes.

I wrote a lot of *Star Wars* fanfic that year. By the age of twelve, I'd graduated to *Conan*. By the age of eighteen, it was Harlan Ellison. By the age of twenty-six, it was Bradbury, by way of Gibson. Today, I hope I write more or less like myself.

Walk the streets of Florence and you'll find a copy of the *David* on practically every corner. For centuries, the way to become a Florentine sculptor has been to copy Michelangelo, to learn from the master. Not just the great Florentine sculptors, either — great or terrible, they all start with the master; it can be the start of a lifelong passion, or a mere fling. The copy can be art, or it can be crap — the best way to find out which kind you've got inside you is to try.

Science fiction has the incredible good fortune to have attracted huge, social groups of fan-fiction writers. Many pros got their start with fanfic (and many of them still work at it in secret), and many fanfic writers are happy to scratch their itch by working only with others' universes, for the sheer joy of it. Some fanfic is great — there's plenty of *Buffy* fanfic that trumps the official, licensed tie-in novels — and some is purely dreadful.

Two things are sure about all fanfic, though: first, that people who write and read fanfic are already avid readers of writers whose work they're paying homage to; and second, that the people who write and read fanfic derive fantastic satisfaction from their labors. This is great news for writers.

Great because fans who are so bought into your fiction that they'll make it their own are fans forever, fans who'll evangelize your work to their friends, fans who'll seek out your work however you publish it.

Great because fans who use your work therapeutically, to work out their own creative urges, are fans who have a damned good reason to stick with the field, to keep on reading even as our numbers dwindle. Even when the fandom revolves around movies or TV shows, fanfic is itself a literary pursuit, something undertaken in the world of words. The fanfic habit is a literary habit.

In Japan, comic book fanfic writers publish fanfic manga called Dōjinshi — some of these titles dwarf the circulation of the work they pay tribute to, and many of them are sold commercially. Japanese comic publishers know a good thing when they see it, and these fanficcers get left alone by the commercial giants they attach themselves to.

And yet for all this, there are many writers who hate fanfic. Some argue that fans have no business appropriating their characters and situations, that it's disrespectful to imagine your precious fictional people in sexual scenarios, or to retell their stories from a different point of view, or to snatch a victorious happy ending from the tragic defeat the writer ended her book with.

Other writers insist that fans who take without asking — or against the writer's wishes — are part of an "entitlement culture" that has decided that it has the moral right to lift scenarios and characters without permission, that this is part of our larger post-modern moral crisis that is making the world a worse place.

Some writers dismiss all fanfic as bad art and therefore unworthy of appropriation. Some call it copyright infringement or trademark infringement, and every now and again, some loony will actually threaten to sue his readers for having had the gall to tell his stories to each other.

I'm frankly flabbergasted by these attitudes. Culture is a lot older than art — that is, we have had social storytelling for a lot longer than we've had a notional class of artistes whose creativity is privileged and elevated to the numinous, far above the everyday creativity of a kid who knows that she can paint and draw, tell a story and sing a song, sculpt and invent a game.

To call this a moral failing — and a *new* moral failing at that! — is to turn your back on millions of years of human history. It's no failing that we internalize the stories we love, that we rework them to suit our minds better. The *Pygmalion* story didn't start with Shaw or the Greeks, nor did it end with *My Fair Lady*. *Pygmalion* is at least thousands of years old — think of Moses passing for the Pharaoh's son! — and has been reworked in a billion bedtime stories, novels, D&D games, movies, fanfic stories, songs, and legends.

Each person who retold *Pygmalion* did something both original — no two tellings are just alike — and derivative, for there are no new ideas under the sun. Ideas are *easy*. Execution is *hard*. That's why writers don't really get excited when they're approached by people with great ideas for novels. We've all got more ideas than we can use — what we lack is the cohesive whole.

Much fanfic — the stuff written for personal consumption or for a small social group — isn't bad art. It's just *not* art. It's not written to make a contribution to the aesthetic development of humanity. It's created to satisfy the deeply human need to play with the stories that constitute our world. There's nothing trivial about telling stories with your friends — even if the stories themselves are trivial. The act of telling stories to one another is practically sacred — and it's unquestionably profound. What's more, lots of retellings are art: witness Pat Murphy's wonderful *There and Back Again* (Tolkien) and Geoff Ryman's brilliant World Fantasy Award-winning *Was* (L. Frank Baum).

The question of respect is, perhaps, a little thornier. The dominant mode of criticism in fanfic circles is to compare a work to the canon — "Would Spock ever say that, in 'real' life?" What's more, fanfic writers will sometimes apply this test to works that are *of* the canon, as in "Spock never would have said that, and Gene Roddenberry has no business telling me otherwise."

This is a curious mix of respect and disrespect. Respect because it's hard to imagine a more respectful stance than the one that says that your work is the yardstick against which all other work is to be measured — what could be more respectful than having your work made into the gold standard? On the other hand, this business of telling writers that they've given their characters the wrong words and deeds can feel obnoxious or insulting.

Writers sometimes speak of their characters running away from them, taking on a life of their own. They say that these characters — drawn from real people in our lives and mixed up with our own imagination — are autonomous pieces of themselves. It's a short leap from there to mystical nonsense about protecting our notional, fictional children from grubby fans who'd set them to screwing each other or bowing and scraping before some thinly veiled version of the fanfic writer herself.

There's something to the idea of the autonomous character. Big chunks of our wetware are devoted to simulating other people, trying to figure out if we are likely to fight or fondle them. It's unsurprising that when you ask your brain to model some other person, it rises to the task. But that's *exactly* what happens to a reader when you hand your book over to him: he simulates your characters in his head, trying to interpret that character's actions through his own lens.

Writers can't ask readers not to interpret their work. You can't enjoy a novel that you haven't interpreted — unless you model the author's characters in your head, you can't care about what they do and why they do it. And once readers model a character, it's only natural that readers will take pleasure in imagining what that character might do offstage, to noodle around with it. This isn't disrespect: it's *active reading*.

Our field is incredibly privileged to have such an active fanfic writing practice. Let's stop treating them like thieves and start treating them like honored guests at a table that we laid just for them.

# Metacrap:
# Putting the Torch to Seven Straw-Men of the Meta-Utopia

(Originally self-published, August 26, 2001.)

ToC:

## 1. Introduction

Metadata is "data about data" — information like keywords, page
length, title, word count, abstract, location, SKU, ISBN, and so
on. Explicit, human-generated metadata has enjoyed recent
trendiness, especially in the world of XML. A typical scenario
goes like this: a number of suppliers get together and agree on
a metadata standard — a Document Type Definition or scheme
— for a given subject area, say washing machines. They agree to
a common vocabulary for describing washing machines: size,
capacity, energy consumption, water consumption, price. They
create machine-readable databases of their inventory, which are
available in whole or in part to search agents and other databases,
so that a consumer can enter the parameters of the washing
machine he's seeking and query multiple sites simultaneously for
an exhaustive list of the available washing machines that meet
his criteria.

If everyone would subscribe to such a system and create good
metadata for the purposes of describing their goods, services,
and information, it would be a trivial matter to search the Internet for highly qualified, context-sensitive results: a fan could
find all the downloadable music in a given genre, a manufacturer
could efficiently discover suppliers, travelers could easily choose
a hotel room for an upcoming trip.

A world of exhaustive, reliable metadata would be a utopia.
It's also a pipe-dream, founded on self-delusion, nerd hubris, and
hysterically inflated market opportunities.

## 2. The problems

There are at least seven insurmountable obstacles between the
world as we know it and meta-utopia. I'll enumerate them below:

## 2.1 People lie

Metadata exists in a competitive world. Suppliers compete to sell their goods, cranks compete to convey their crackpot theories (mea culpa), artists compete for audience. Attention spans and wallets may not be zero-sum, but they're damned close.

That's why:

> A search for any commonly referenced term at a search-engine like AltaVista will often turn up at least one porn link in the first ten results.

> Your mailbox is full of spam with subject lines like "Re: The information you requested."

> Publishers Clearing House sends out advertisements that holler "You may already be a winner!"

> Press-releases have gargantuan lists of empty buzzwords attached to them.

Meta-utopia is a world of reliable metadata. When poisoning the well confers benefits to the poisoners, the meta-waters get awfully toxic in short order.

## 2.2 People are lazy

You and I are engaged in the incredibly serious business of creating information. Here in the Info-Ivory-Tower, we understand the importance of creating and maintaining excellent metadata for our information.

But info-civilians are remarkably cavalier about their information. Your clueless aunt sends you email with no subject line, half the pages on Geocities are called "Please title this page," and your boss stores all of his files on his desktop with helpful titles like "UNTITLED.DOC."

This laziness is bottomless. No amount of ease-of-use will end it. To understand the true depths of meta-laziness, download ten random MP3 files from Napster. Chances are, at least one will have no title, artist, or track information — this despite the fact that adding in this info merely requires clicking the "Fetch Track Info from CDDB" button on every MP3-ripping application.

Short of breaking fingers or sending out squads of vengeful info-ninjas to add metadata to the average user's files, we're never gonna get there.

## 2.3 People are stupid

Even when there's a positive benefit to creating good metadata, people steadfastly refuse to exercise care and diligence in their metadata creation.

Take eBay: every seller there has a damned good reason for double-checking their listings for typos and misspellings. Try searching for "plam" on eBay. Right now, that turns up *nine* typoed listings for "Plam Pilots." Misspelled listings don't show up in correctly spelled searches and hence garner fewer bids and lower sale-prices. You can almost always get a bargain on a Plam Pilot at eBay.

The fine (and gross) points of literacy — spelling, punctuation, grammar — elude the vast majority of the Internet's users. To believe that J. Random Users will suddenly and *en masse* learn to spell and punctuate — let alone accurately categorize their information according to whatever hierarchy they're supposed to be using — is self-delusion of the first water.

## 2.4 Mission: Impossible — know thyself

In meta-utopia, everyone engaged in the heady business of describing stuff carefully weighs the stuff in the balance and accurately divines the stuff's properties, noting those results.

Simple observation demonstrates the fallacy of this assumption. When Nielsen used log-books to gather information on the viewing habits of their sample families, the results were heavily skewed to *Masterpiece Theater* and *Sesame Street*. Replacing the journals with set-top boxes that reported what the set was actually tuned to showed what the average American family was really watching: naked midget wrestling, *America's Funniest Botched Cosmetic Surgeries* and Jerry Springer presents: "My daughter dresses like a slut!"

Ask a programmer how long it'll take to write a given module, or a contractor how long it'll take to fix your roof. Ask a laconic Southerner how far it is to the creek. Better yet, throw darts — the answer's likely to be just as reliable.

People are lousy observers of their own behaviors. Entire religions are formed with the goal of helping people understand themselves better; therapists rake in billions working for this very end.

Why should we believe that using metadata will help J. Random User get in touch with her Buddha nature?

## 2.5 Schemas aren't neutral

In meta-utopia, the lab-coated guardians of epistemology sit down and rationally map out a hierarchy of ideas, something like this:

> Nothing:
> > Black holes
>
> Everything:
> > Matter:

Earth:
    Planets
    Washing machines
Wind:
    Oxygen
    Poo-gas
Fire:
    Nuclear fission
    Nuclear fusion
    "Mean Devil Woman" Louisiana Hot-Sauce

In a given sub-domain, say, Washing machines, experts agree on sub-hierarchies, with classes for reliability, energy consumption, color, size, etc.

This presumes that there is a "correct" way of categorizing ideas, and that reasonable people, given enough time and incentive, can agree on the proper means for building a hierarchy.

Nothing could be farther from the truth. Any hierarchy of ideas necessarily implies the importance of some axes over others. A manufacturer of small, environmentally conscious washing machines would draw a hierarchy that looks like this:

Energy consumption:
    Water consumption:
        Size:
            Capacity:
                Reliability

While a manufacturer of glitzy, feature-laden washing machines would want something like this:

Color:
  Size:
    Programmability:
      Reliability

The conceit that competing interests can come to an easy accord on a common vocabulary totally ignores the power of organizing principles in a marketplace.

## 2.6 Metrics influence results

Agreeing to a common yardstick for measuring the important stuff in any domain necessarily privileges the items that score high on that metric, regardless of those items' overall suitability. IQ tests privilege people who are good at IQ tests, Nielsen Ratings privilege 30- and 60-minute TV shows (which is why MTV doesn't show videos any more — Nielsen couldn't generate ratings for 3-minute mini-programs, and so MTV couldn't demonstrate the value of advertising on its network), raw megahertz scores privilege Intel's CISC chips over Motorola's RISC chips.

Ranking axes are mutually exclusive: software that scores high for security scores low for convenience, desserts that score high for decadence score low for healthiness. Every player in a metadata standards body wants to emphasize their high-scoring axes and de-emphasize (or, if possible, ignore altogether) their low-scoring axes.

It's wishful thinking to believe that a group of people competing to advance their agendas will be universally pleased with any hierarchy of knowledge. The best that we can hope for is a *detente* in which everyone is equally miserable.

*2.7 There's more than one way to describe something*

"No, I'm not watching cartoons! It's cultural anthropology."

"This isn't smut, it's *art*."

"It's not a bald spot, it's a *solar panel for a sex-machine*."

Reasonable people can disagree forever on how to describe something. Arguably, your Self is the collection of associations and descriptors you ascribe to ideas. Requiring everyone to use the same vocabulary to describe their material denudes the cognitive landscape, enforces homogeneity in ideas.

And that's just not right.

## 3. Reliable metadata

Do we throw out metadata, then?

Of course not. Metadata can be quite useful, if taken with a sufficiently large pinch of salt. The meta-utopia will never come into being, but metadata is often a good means of making rough assumptions about the information that floats through the Internet.

Certain kinds of implicit metadata are awfully useful, in fact. Google exploits metadata about the structure of the World Wide Web: by examining the number of links pointing at a page (and the number of links pointing at each linker), Google can derive statistics about the number of Web-authors who believe that that page is important enough to link to, and hence make extremely reliable guesses about how reputable the information on that page is.

This sort of observational metadata is far more reliable than the stuff that human beings create for the purposes of having their documents found. It cuts through the marketing bullshit, the self-delusion, and the vocabulary collisions.

Taken more broadly, this kind of metadata can be thought of as a pedigree: Who thinks that this document is valuable? How closely correlated have this person's value judgments been with mine in times gone by? This kind of implicit endorsement of information is a far better candidate for an information-retrieval panacea than all the world's schema combined.

# Amish for QWERTY

(Originally published on the O'Reilly Network [*www.oreillynet.com*], 07/09/2003)

I learned to type before I learned to write. The QWERTY keyboard layout is hard-wired to my brain, such that I can't write anything of significance without that I have a 101-key keyboard in front of me. This has always been a badge of geek pride: unlike the creaking pen-and-ink dinosaurs that I grew up reading, I'm well adapted to the modern reality of technology. There's a secret elitist pride in touch-typing on a laptop while staring off into space, fingers flourishing and caressing the keys.

But last week, my pride got pricked. I was brung low by a phone. Some very nice people from Nokia loaned me a very latest-and-greatest camera-phone, the kind of gadget I've described in my science fiction stories. As I prodded at the little 12-key interface, I felt like my father, a 60s-vintage computer scientist who can't get his wireless network to work, must feel. Like a creaking dino. Like history was passing me by. I'm 31, and I'm obsolete. Or at least Amish.

People think the Amish are technophobes. Far from it. They're ideologues. They have a concept of what right-living consists of, and they'll use any technology that serves that ideal — and mercilessly eschew any technology that would subvert it. There's nothing wrong with driving the wagon to the next farm when you want to hear from your son, so there's no need to put a phone in the kitchen. On the other hand, there's nothing *right* about your livestock dying for lack of care, so a cellphone that can call the veterinarian can certainly find a home in the horse barn.

For me, right-living is the 101-key, QWERTY, computer-centric mediated lifestyle. It's having a bulky laptop in my bag, crouching by the toilets at a strange airport with my AC adapter plugged into the always-awkwardly-placed power source, running software that I chose and installed, communicating over the wireless network. I use a network that has no incremental cost for communication, and a device that lets me install any software without permission from anyone else. Right-living is the highly mutated, commodity-hardware-based, public and free Internet. I'm QWERTY-Amish, in other words.

I'm the kind of perennial early adopter who would gladly volunteer to beta test a neural interface, but I find myself in a moral panic when confronted with the 12-button keypad on a cellie, even though that interface is one that has been greedily adopted by billions of people worldwide, from strap-hanging Japanese schoolgirls to Kenyan electoral scrutineers to Filipino guerrillas in the bush. The idea of paying for every message makes my hackles tumesce and evokes a reflexive moral conviction that text-messaging is inherently undemocratic, at least compared to free-as-air email. The idea of only running the software that big-brother telco has permitted me on my handset makes me want to run for the hills.

The thumb-generation who can tap out a text-message under their desks while taking notes with the other hand — they're in for it, too. The pace of accelerated change means that we're all of us becoming wed to interfaces — ways of communicating with our tools and our world — that are doomed, doomed, doomed. The 12-buttoners are marrying the phone company, marrying a centrally controlled network that requires permission to use and improve, a Stalinist technology whose centralized choke points are subject to regulation and the vagaries of the telcos. Long after the phone companies have been out-competed by the pure and open Internet (if such a glorious day comes to pass), the kids of today will be bound by its interface and its conventions.

The sole certainty about the future is its Amishness. We will all bend our brains to suit an interface that we will either have to abandon or be left behind. Choose your interface — and the values it implies — carefully, then, before you wed your thought processes to your fingers' dance. It may be the one you're stuck with.

# Ebooks: Neither E, Nor Books

(Originally given as a paper at the O'Reilly Emerging Technology Conference, San Diego, California, February 12, 2004.)

**Forematter:**

This talk was initially given at the O'Reilly Emerging Technology Conference [*http://conferences.oreillynet.com/et2004/*], along with a set of slides that, for copyright reasons (ironic!) can't be released alongside of this file. However, you will find, interspersed in this text, notations describing the places where new slides should be loaded, in [square-brackets].

For starters, let me try to summarize the lessons and intuitions I've had about ebooks from my release of two novels and most of a short story collection online under a Creative Commons license. A parodist who published a list of alternate titles for the presentations at this event called this talk "eBooks Suck Right Now," [eBooks suck right now] and as funny as that is, I don't think it's true.

No, if I had to come up with another title for this talk, I'd call it: "Ebooks: You're Soaking in Them." [Ebooks: You're Soaking in Them] That's because I think that the shape of ebooks to come is almost visible in the way that people interact with text today, and that the job of authors who want to become rich and famous is to come to a better understanding of that shape.

I haven't come to a perfect understanding. I don't know what the future of the book looks like. But I have ideas, and I'll share them with you:

1. Ebooks aren't marketing. [Ebooks aren't marketing] OK, so ebooks *are* marketing: that is to say that giving away ebooks sells more books. Baen Books, who do a lot of series publishing, have found that giving away electronic editions of the previous installments in their series to coincide with the release of a new volume sells the hell out of the new book — and the backlist. And the number of people who wrote to me to tell me about how much they dug the ebook and so bought the paper book far exceeds the number of people who wrote to me and said, "Ha, ha, you hippie, I read your book for free and now I'm not gonna buy it." But ebooks *shouldn't* be just about marketing: ebooks are a goal unto themselves. In the final analysis, more people will read more words off more screens and fewer words off fewer pages and when those two lines cross, ebooks are gonna have to be the way that writers earn their keep, not the way that they promote the dead-tree editions.

2. Ebooks complement paper books. [Ebooks complement paper books] Having an ebook is good. Having a paper book is good. Having both is even better. One reader wrote to me and said that he read half my first novel from the bound book, and printed the other half on scrap-paper to read at the beach. Students write to me to say that it's easier to do their term papers if they can copy and paste their quotations into their word-processors. Baen readers use the electronic editions of their favorite series to build concordances of characters, places, and events.

3. Unless you own the ebook, you don't 0wn the book [Unless you own the ebook, you don't 0wn the book]. I take the view that

the book is a "practice" — a collection of social and economic and artistic activities — and not an "object." Viewing the book as a "practice" instead of an object is a pretty radical notion, and it begs the question: Just what the hell is a book? Good question. I write all of my books in a text-editor [TEXT EDITOR SCREENGRAB] (BBEdit, from Barebones Software — as fine a text-editor as I could hope for). From there, I can convert them into a formatted two-column PDF [TWO-UP SCREENGRAB]. I can turn them into an HTML file [BROWSER SCREENGRAB]. I can turn them over to my publisher, who can turn them into galleys, advanced review copies, hardcovers, and paperbacks. I can turn them over to my readers, who can convert them to a bewildering array of formats [DOWNLOAD PAGE SCREENGRAB]. Brewster Kahle's Internet Bookmobile can convert a digital book into a four-color, full-bleed, perfect-bound, laminated-cover, printed-spine paper book in ten minutes, for about a dollar. Try converting a paper book to a PDF or an html file or a text file or a Rocketbook or a printout for a buck in ten minutes! It's ironic, because one of the frequently cited reasons for preferring paper to ebooks is that paper books confer a sense of ownership of a physical object. Before the dust settles on this ebook thing, owning a paper book is going to feel less like ownership than having an open digital edition of the text.

4. Ebooks are a better deal for writers. [Ebooks are a better deal for writers] The compensation for writers is pretty thin on the ground. *Amazing Stories*, Hugo Gernsback's original science fiction magazine, paid a couple cents a word. Today, science fiction magazines pay...a couple cents a word. The sums involved are so minuscule, they're not even insulting: they're *quaint* and *historical*, like the WHISKEY 5 CENTS sign over the bar at a pioneer village. Some writers do make it big, but they're *rounding*

*errors* as compared to the total population of sf writers earning some of their living at the trade. Almost all of us could be making more money elsewhere (though we may dream of earning a stephenkingload of money, and of course, no one would play the lotto if there were no winners). The primary incentive for writing has to be artistic satisfaction, egoboo, and a desire for posterity. Ebooks get you that. Ebooks become a part of the corpus of human knowledge because they get indexed by search engines and replicated by the hundreds, thousands, or millions. They can be googled.

Even better: they level the playing field between writers and trolls. When Amazon kicked off, many writers got their knickers in a tight and powerful knot at the idea that axe-grinding yahoos were filling the Amazon message-boards with ill-considered slams at their work — for, if a personal recommendation is the best way to sell a book, then certainly a personal condemnation is the best way to *not* sell a book. Today, the trolls are still with us, but now, the readers get to decide for themselves. Here's a bit of a review of *Down and Out in the Magic Kingdom* that was recently posted to Amazon by "A reader from Redwood City, CA":

[QUOTED TEXT]
> I am really not sure what kind of drugs critics are
> smoking, or what kind of payola may be involved. But
> regardless of what *Entertainment Weekly* says, whatever
> this newspaper or that magazine says, you shouldn't
> waste your money. Download it for free from Corey's
> (sic) site, read the first page, and look away in
> disgust — this book is for people who think Dan
> Brown's *Da Vinci Code* is great writing.

Back in the old days, this kind of thing would have really

pissed me off. Axe-grinding, mouth-breathing yahoos defaming my good name! My stars and mittens! But take a closer look at that damning passage:

[PULL-QUOTE]
> Download it for free from Corey's site, read the first
> page

You see that? Hell, this guy is *working for me!* [ADDITIONAL PULL QUOTES] Someone accuses a writer I'm thinking of reading of paying off *Entertainment Weekly* to say nice things about his novel, "a surprisingly bad writer," no less, whose writing is "stiff, amateurish, and uninspired!" I wanna check that writer out. And I can. In one click. And then I can make up my own mind.

You don't get far in the arts without healthy doses of both ego and insecurity, and the downside of being able to google up all the things that people are saying about your book is that it can play right into your insecurities — "all these people will have it in their minds not to bother with my book because they've read the negative interweb reviews!" But the flipside of that is the ego: "If only they'd give it a shot, they'd see how good it is." And the more scathing the review is, the more likely they are to give it a shot. Any press is good press, so long as they spell your URL right (and even if they spell your name wrong!).

5. Ebooks need to embrace their nature. [Ebooks need to embrace their nature] The distinctive value of ebooks is orthogonal to the value of paper books, and it revolves around the mix-ability and send-ability of electronic text. The more you constrain an ebook's distinctive value propositions — that is, the more you restrict a reader's ability to copy, transport, or transform an ebook — the more it has to be valued on the same axes as a paper book. Ebooks

*fail* on those axes. Ebooks don't beat paper books for sophisticated typography, they can't match them for quality of paper or the smell of the glue. But just try sending a paper book to a friend in Brazil, for free, in less than a second. Or loading a thousand paper books into a little stick of flash-memory dangling from your keychain. Or searching a paper book for every instance of a character's name to find a beloved passage. Hell, try clipping a pithy passage out of a paper book and pasting it into your sig-file.

6. Ebooks demand a different attention span (but not a shorter one). [Ebooks demand a different attention span (but not a shorter one)] Artists are always disappointed by their audience's attention spans. Go back far enough and you'll find cuneiform etchings bemoaning the current Sumerian go-go lifestyle with its insistence on myths with plotlines and characters and action, not like we had in the old days. As artists, it would be a hell of a lot easier if our audiences were more tolerant of our penchant for boring them. We'd get to explore a lot more ideas without worrying about tarting them up with easy-to-swallow chocolate coatings of entertainment. We like to think of shortened attention spans as a product of the information age, but check this out:

[Nietzsche quote]
> To be sure one thing necessary above all: if one is to
> practice reading as an *art* in this way, something
> needs to be un-learned most thoroughly in these days.

In other words, if my book is too boring, it's because you're not paying enough attention. Writers say this stuff all the time, but this quote isn't from this century or the last. [Nietzsche quote with attribution] It's from the preface to Nietzsche's *On the Genealogy of Morality,* published in 1887.

Yeah, our attention spans are *different* today, but they aren't necessarily *shorter*. Warren Ellis's fans managed to hold the story-line for *Transmetropolitan* [*Transmet* cover] in their minds for *five years* while the story trickled out in monthly funnybook installments. J. K. Rowlings' installments on the Harry Potter series get fatter and fatter with each new volume. Entire forests are sacrificed to long-running series fiction like Robert Jordan's Wheel of Time books, each of which is approximately 20,000 pages long (I may be off by an order of magnitude one way or another here). Sure, presidential debates are conducted in soundbites today and not the days-long oratory extravaganzas of the Lincoln-Douglas debates, but people manage to pay attention to the 24-month-long presidential campaigns from start to finish.

7. We need *all* the ebooks. [We need *all* the ebooks] The vast majority of the words ever penned are lost to posterity. No one library collects all the still-extant books ever written and no one person could hope to make a dent in that corpus of written work. None of us will ever read more than the tiniest sliver of human literature. But that doesn't mean that we can stick with just the most popular texts and get a proper ebook revolution.

For starters, we're all edge-cases. Sure, we all have the shared desire for the core canon of literature, but each of us wants to complete that collection with different texts that are as distinctive and individualistic as fingerprints. If we all look like we're doing the same thing when we read, or listen to music, or hang out in a chatroom, that's because we're not looking closely enough. The shared-ness of our experience is only present at a coarse level of measurement: once you get into really granular observation, there are as many differences in our "shared" experience as there are similarities.

More than that, though, is the way that a large collection of electronic text differs from a small one: it's the difference between a single book, a shelf full of books, and a library of books. Scale makes things different. Take the Web: none of us can hope to read even a fraction of all the pages on the Web, but by analyzing the link structures that bind all those pages together, Google is able to actually tease out machine-generated conclusions about the relative relevance of different pages to different queries. None of us will ever eat the whole corpus, but Google can digest it for us and excrete the steaming nuggets of goodness that make it the search-engine miracle it is today.

8. Ebooks are like paper books. [Ebooks are like paper books] To round out this talk, I'd like to go over the ways that ebooks are more like paper books than you'd expect. One of the truisms of retail theory is that purchasers need to come into contact with a good several times before they buy — seven contacts is tossed around as the magic number. That means that my readers have to hear the title, see the cover, pick up the book, read a review, and so forth, seven times, on average, before they're ready to buy.

There's a temptation to view downloading a book as comparable to bringing it home from the store, but that's the wrong metaphor. Some of the time, maybe most of the time, downloading the text of the book is like taking it off the shelf at the store and looking at the cover and reading the blurbs (with the advantage of not having to come into contact with the residual DNA and burger king left behind by everyone else who browsed the book before you). Some writers are horrified at the idea that three hundred thousand copies of my first novel were downloaded and "only" ten thousand or so were sold so far. If it were the case that for every copy sold, thirty were taken home from the store, that would be a horrifying outcome, for sure. But look at it another way: if one out of every thirty people who glanced at the cover of my book bought it, I'd be a happy author. And I am. Those downloads cost me no more than glances at the cover in a bookstore, and the sales are healthy.

We also like to think of physical books as being inherently *countable* in a way that digital books aren't (an irony, since computers are damned good at counting things!). This is important, because writers get paid on the basis of the number of copies of their books that sell, so having a good count makes a difference. And indeed, my royalty statements contain precise numbers for copies printed, shipped, returned, and sold.

But that's a false precision. When the printer does a run of a book, it always runs a few extra at the start and finish of the run to make sure that the setup is right and to account for the occasional rip, drop, or spill. The actual total number of books printed is approximately the number of books ordered, but never exactly — if you've ever ordered 500 wedding invitations, chances are you received 500-and-a-few back from the printer and that's why.

And the numbers just get fuzzier from there. Copies are stolen. Copies are dropped. Shipping people get the count wrong. Some copies end up in the wrong box and go to a bookstore that didn't order them and isn't invoiced for them and end up on a sale table or in the trash. Some copies are returned as damaged. Some are returned as unsold. Some come back to the store the next morning accompanied by a whack of buyer's remorse. Some go to the place where the spare sock in the dryer ends up.

The numbers on a royalty statement are actuarial, not actual. They represent a kind of best-guess approximation of the copies shipped, sold, returned, and so forth. Actuarial accounting works pretty well: well enough to run the juggernaut banking, insurance, and gambling industries on. It's good enough for divvying up the royalties paid by musical rights societies for radio airplay and live performance. And it's good enough for counting how many copies of a book are distributed online or off.

Counts of paper books are differently precise from counts of electronic books, sure: but neither one is inherently countable.

And finally, of course, there's the matter of selling books. However an author earns her living from her words, printed or encoded, she has as her first and hardest task to find her audience. There are more competitors for our attention than we can possibly reconcile, prioritize, or make sense of. Getting a book under the right person's nose, with the right pitch, is the hardest and most important task any writer faces.

I care about books, a lot. I started working in libraries and bookstores at the age of 12 and kept at it for a decade, until I was lured away by the siren song of the tech world. I knew I wanted to be a writer at the age of 12, and now, 20 years later, I have three novels, a short story collection, and a nonfiction book out, two more novels under contract, and another book in the works. [BOOK COVERS] I've won a major award in my genre, science fiction [CAMPBELL AWARD], and I'm nominated for another one, the 2003 Nebula Award for best novelette. [NEBULA]

I own a *lot* of books. Easily more than 10,000 of them, in storage on both coasts of the North American continent [LIBRARY LADDER]. I have to own them, since they're the tools of my trade: the reference works I refer to as a novelist and writer today. Most of the literature I dig is very short-lived, it disappears from the shelf after just a few months, usually for good. Science fiction is inherently ephemeral. [ACE DOUBLES]

Now, as much as I love books, I love computers, too. Computers are fundamentally different from modern books in the same way that printed books are different from monastic Bibles: they are malleable. Time was, a "book" was something produced by many months' labor by a scribe, usually a monk, on some kind of durable and sexy substrate like foetal lambskin. [ILLUMINATED BIBLE] Gutenberg's Xerox machine changed all that, changed a book into something that could be simply run off a press in a few minutes' time, on substrate more suitable to ass-wiping than exaltation in a place of honor in the cathedral. The Gutenberg press meant that rather than owning one or two books, a member of the ruling class could amass a library, and that rather than picking only a few subjects from enshrinement in print, a huge variety of subjects could be addressed on paper and handed from person to person. [KAPITAL/TIJUANA BIBLE]

Most new ideas start with a precious few certainties and a lot of speculation. I've been doing a bunch of digging for certainties and a lot of speculating lately, and the purpose of this talk is to lay out both categories of ideas.

This all starts with my first novel, *Down and Out in the Magic Kingdom* [COVER], which came out on January 9, 2003. At that time, there was a lot of talk in my professional circles about, on the one hand, the dismal failure of ebooks, and, on the other, the new and scary practice of ebook "piracy." [alt.binaries.ebooks screengrab] It was strikingly weird that no one seemed to notice that the idea of ebooks as a "failure" was at strong odds with the notion that electronic book "piracy" was worth worrying about: I mean, if ebooks are a failure, then who gives a rats if intarweb dweebs are trading them on Usenet?

A brief digression here, on the double meaning of "ebooks." One meaning for that word is "legitimate" ebook ventures, that is to say, rightsholder-authorized editions of the texts of books, released in a proprietary, use-restricted format, sometimes for use on a general-purpose PC and sometimes for use on a special-purpose hardware device like the NuvoMedia Rocketbook [ROCKETBOOK]. The other meaning for ebook is a "pirate" or unauthorized electronic edition of a book, usually made by cutting the binding off of a book and scanning it a page at a time, then running the resulting bitmaps through an optical character recognition app to convert them into ASCII text, to be cleaned up by hand. These books are pretty buggy, full of errors introduced by the OCR. A lot of my colleagues worry that these books also have deliberate errors, created by mischievous book-rippers who cut, add, or change text in order to "improve" the work. Frankly, I have never seen any evidence that any book-ripper is interested in doing this, and until I do, I think that this is the last thing anyone should be worrying about.

Back to *Down and Out in the Magic Kingdom* [COVER]. Well, not yet. I want to convey to you the depth of the panic in my field over ebook piracy, or "bookwarez" as it is known in book-ripper circles. Writers were joining the discussion on alt.binaries .ebooks using assumed names, claiming fear of retaliation from scary hax0r kids who would presumably screw up their credit-ratings in retaliation for being called thieves. My editor, a blogger, hacker, and guy-in-charge-of-the-largest-sf-line-in-the-world named Patrick Nielsen Hayden, posted to one of the threads in the newsgroup, saying, in part [SCREENGRAB]:

> Pirating copyrighted etext on Usenet and elsewhere
> is going to happen more and more, for the same
> reasons that everyday folks make audio cassettes from
> vinyl LPs and audio CDs, and videocassette copies
> store-bought videotapes. Partly it's greed; partly it's
> annoyance over retail prices; partly it's the desire to
> Share Cool Stuff (a motivation usually underrated by
> the victims of this kind of small-time hand-level piracy).
> Instantly going to Defcon One over it and claiming
> it's morally tantamount to mugging little old ladies
> in the street will make it kind of difficult to move

> forward from that position when it doesn't work.
> In the 1970s, the record industry shrieked that
> "home taping is killing music." It's hard for ordinary
> folks to avoid noticing that music didn't die. But the
> record industry's credibility on the subject wasn't
> exactly enhanced.

Patrick and I have a long relationship, starting when I was 18 years old and he kicked in toward a scholarship fund to send me to a writers' workshop, continuing to a fateful lunch in New York in the mid-nineties when I showed him a bunch of Project Gutenberg texts on my Palm Pilot and inspired him to start licensing Tor's titles for PDAs [PEANUTPRESS SCREENGRAB], to the turn-of-the-millennium when he bought and then published my first novel (he's bought three more since — I really like Patrick!).

Right as bookwarez newsgroups were taking off, I was shocked silly by legal action by one of my colleagues against AOL/Time-Warner for carrying the alt.binaries.ebooks newsgroup. This writer alleged that AOL should have a duty to remove this newsgroup, since it carried so many infringing files, and that its failure to do so made it a contributory infringer, and so liable for the incredibly stiff penalties afforded by our newly minted copyright laws like the No Electronic Theft Act and the loathsome Digital Millennium Copyright Act or DMCA.

Now there was a scary thought: there were people out there who thought the world would be a better place if ISPs were given the duty of actively policing and censoring the websites and newsfeeds their customers had access to, including a requirement that ISPs needed to determine, all on their own, what was an unlawful copyright infringement — something more usually left up to judges in the light of extensive amicus briefings from esteemed copyright scholars [WIND DONE GONE GRAPHIC].

This was a stupendously dumb idea, and it offended me down to my boots. Writers are supposed to be advocates of free expression, not censorship. It seemed that some of my colleagues loved the First Amendment, but they were reluctant to share it with the rest of the world.

Well, dammit, I had a book coming out, and it seemed to be an opportunity to try to figure out a little more about this ebook stuff. On the one hand, ebooks were a dismal failure. On the other hand, there were more books posted to alt.binaries.ebooks every day.

This leads me into the two certainties I have about ebooks:

1. More people are reading more words off more screens every day [GRAPHIC].
2. Fewer people are reading fewer words off fewer pages every day [GRAPHIC].

These two certainties begged a lot of questions.

[CHART: EBOOK FAILINGS]

* Screen resolutions are too low to effectively replace paper.
* People want to own physical books because of their visceral appeal (often this is accompanied by a little sermonette on how good books smell, or how good they look on a bookshelf, or how evocative an old curry stain in the margin can be).
* You can't take your ebook into the tub.
* You can't read an ebook without power and a computer.
* File-formats go obsolete, paper has lasted for a long time.

None of these seemed like very good explanations for the "failure" of ebooks to me. If screen resolutions are too low to replace paper, then how come everyone I know spends more time reading off a screen every year, up to and including my sainted grandmother (geeks have a really crappy tendency to argue that certain technologies aren't ready for primetime because their grandmothers won't use them — well, my grandmother sends me email all the time. She types 70 words per minute, and loves to show off grandsonular email to her pals around the pool at her Florida retirement condo).

The other arguments were a lot more interesting, though. It seemed to me that electronic books are *different* from paper books, and have different virtues and failings. Let's think a little about what the book has gone through in years gone by. This is interesting because the history of the book is the history of the Enlightenment, the Reformation, the Pilgrims, and, ultimately, the colonizing of the Americas and the American Revolution.

Broadly speaking, there was a time when books were hand-printed on rare leather by monks. The only people who could read them were priests, who got a regular eyeful of the really cool cartoons the monks drew in the margins. The priests read the books aloud, in Latin [LATIN BIBLE] (to a predominantly non-Latin-speaking audience) in cathedrals, wreathed in pricey incense that rose from censers swung by altar boys.

Then Johannes Gutenberg invented the printing press. Martin Luther turned that press into a revolution. [LUTHER BIBLE] He printed Bibles in languages that non-priests could read, and distributed them to normal people who got to read the word of God all on their own. The rest, as they say, is history.

Here are some interesting things to note about the advent of the printing press:

[CHART: LUTHER VERSUS THE MONKS]

* Luther Bibles lacked the manufacturing quality of the illuminated Bibles. They were comparatively cheap and lacked the typographical expressiveness that a really talented monk could bring to bear when writing out the word of God.

* Luther Bibles were utterly unsuited to the traditional use-case for Bibles. A good Bible was supposed to reinforce the authority of the man at the pulpit. It needed heft, it needed impressiveness, and most of all, it needed rarity.

* The user-experience of Luther Bibles sucked. There was no incense, no altar boys, and who (apart from the priesthood) knew that reading was so friggin' hard on the eyes?

* Luther Bibles were a lot less trustworthy than the illuminated numbers. Anyone with a press could run one off, subbing in any apocryphal text he wanted — and who knew how accurate that translation was? Monks had an entire Papacy behind them, running a quality-assurance operation that had stood Europe in good stead for centuries.

In the late nineties, I went to conferences where music execs patiently explained that Napster was doomed, because you didn't get any cover-art or liner-notes with it, you couldn't know if the rip was any good, and sometimes the connection would drop mid-download. I'm sure that many cardinals espoused the points raised above with equal certainty.

What the record execs and the cardinals missed was all the ways that Luther Bibles kicked ass:

[CHART: WHY LUTHER BIBLES KICKED ASS]

* They were cheap and fast. Loads of people could acquire them without having to subject themselves to the authority and approval of the Church.

* They were in languages that non-priests could read. You no longer had to take the Church's word for it when its priests explained what God really meant.

* They birthed a printing-press ecosystem in which lots of books flourished. New kinds of fiction, poetry, politics, scholarship, and so on were all enabled by the printing presses whose initial popularity was spurred by Luther's ideas about religion.

Note that all of these virtues are orthogonal to the virtues of a monkish Bible. That is, none of the things that made the Gutenberg press a success were the things that made monk-Bibles a success.

By the same token, the reasons to love ebooks have precious little to do with the reasons to love paper books.

[CHART: WHY EBOOKS KICK ASS]

* They are easy to share. *Divine Secrets of the Ya-Ya Sisterhood* went from a midlist title to a bestseller by being passed from hand to hand by women in reading circles. Slashdorks and other netizens have social lives as rich as reading-circlites, but they don't ever get to see each other face to face; the only kind of book they can pass from hand to hand is an ebook. What's more, the single factor most correlated with a purchase is a recommendation from a friend — getting a book recommended by a pal is more likely to sell you on it than having read and enjoyed the preceding volume in a series!

* They are easy to slice and dice. This is where the Mac evangelist in me comes out — minority platforms matter. It's a truism of the Napsterverse that most of the files downloaded are bog-standard top-40 tracks, like 90 percent or so, and I believe it. We all want to hear popular music. That's why it's popular. But

the interesting thing is the other 10 percent. Bill Gates told the *New York Times* that Microsoft lost the search wars by doing "a good job on the 80 percent of common queries and ignor[ing] the other stuff. But it's the remaining 20 percent that counts, because that's where the quality perception is." Why did Napster captivate so many of us? Not because it could get us the top-40 tracks that we could hear just by snapping on the radio: it was because 80 percent of the music ever recorded wasn't available for sale anywhere in the world, and in that 80 percent were all the songs that had ever touched us, all the earworms that had been lodged in our hindbrains, all the stuff that made us smile when we heard it. Those songs are different for all of us, but they share the trait of making the difference between a compelling service and, well, top-40 Clear Channel radio programming. It was the minority of tracks that appealed to the majority of us. By the same token, the malleability of electronic text means that it can be readily repurposed: you can throw it on a webserver or convert it to a format for your favorite PDA; you can ask your computer to read it aloud or you can search the text for a quotation to cite in a book report or to use in your sig. In other words, most people who download the book do so for the predictable reason, and in a predictable format — say, to sample a chapter in the HTML format before deciding whether to buy the book — but the thing that differentiates a boring etext experience from an exciting one is the minority use — printing out a couple chapters of the book to bring to the beach rather than risk getting the hardcopy wet and salty.

Toolmakers and software designers are increasingly aware of the notion of "affordances" in design. You can bash a nail into the wall with any heavy, heftable object from a rock to a hammer to a cast-iron skillet. However, there's something about a hammer that cries out for nail-bashing, it has affordances that tilt its holder towards swinging it. And, as we all know, when all you have is a hammer, everything starts to look like a nail.

The affordance of a computer — the thing it's designed to do — is to slice-and-dice collections of bits. The affordance of the Internet is to move bits at very high speed around the world at little-to-no cost. It follows from this that the center of the ebook experience is going to involve slicing and dicing text and sending it around.

Copyright lawyers have a word for these activities: infringement. That's because copyright gives creators a near-total monopoly over copying and remixing of their work, pretty much forever (theoretically, copyright expires, but in actual practice, copyright gets extended every time the early Mickey Mouse cartoons are about to enter the public domain, because Disney swings a very big stick on the Hill).

This is a huge problem. The biggest possible problem. Here's why:

[CHART: HOW BROKEN COPYRIGHT SCREWS EVERYONE]

* Authors freak out. Authors have been schooled by their peers that strong copyright is the only thing that keeps them from getting savagely rogered in the marketplace. This is pretty much true: it's strong copyright that often defends authors from their publishers' worst excesses. However, it doesn't follow that strong copyright protects you from your *readers*.

* Readers get indignant over being called crooks. Seriously. You're a small businessperson. Readers are your customers. Calling them crooks is bad for business.

* Publishers freak out. Publishers freak out, because they're in the business of grabbing as much copyright as they can and hanging onto it for dear life because, dammit, you never know. This is why science fiction magazines try to trick writers into sign-

ing over improbable rights for things like theme park rides and action figures based on their work — it's also why literary agents are now asking for copyright-long commissions on the books they represent: copyright covers so much ground and takes too long to shake off, who wouldn't want a piece of it?

* Liability goes through the roof. Copyright infringement, especially on the Net, is a supercrime. It carries penalties of $150,000 per infringement, and aggrieved rightsholders and their representatives have all kinds of special powers, like the ability to force an ISP to turn over your personal information before showing evidence of your alleged infringement to a judge. This means that anyone who suspects that he might be on the wrong side of copyright law is going to be terribly risk-averse: publishers non-negotiably force their authors to indemnify them from infringement claims and go one better, forcing writers to prove that they have "cleared" any material they quote, even in the case of brief fair-use quotations, like song titles at the opening of chapters. The result is that authors end up assuming potentially life-destroying liability, are chilled from quoting material around them, and are scared off of public domain texts because an honest mistake about the public-domain status of a work carries such a terrible price.

* Posterity vanishes. In the *Eldred v. Ashcroft* Supreme Court hearing last year, the court found that 98 percent of the works in copyright are no longer earning money for anyone, but that figuring out who these old works belong to with the degree of certainty that you'd want when one mistake means total economic apocalypse would cost more than you could ever possibly earn on them. That means that 98 percent of works will largely expire long before the copyright on them does. Today, the names of science fiction's ancestral founders — Mary Shelley, Arthur Conan Doyle, Edgar Allan Poe, Jules Verne, H. G. Wells — are still known, their

work still a part of the discourse. Their spiritual descendants from Hugo Gernsback onward may not be so lucky — if their work continues to be "protected" by copyright, it might just vanish from the face of the earth before it reverts to the public domain.

This isn't to say that copyright is bad, but that there's such a thing as good copyright and bad copyright, and that sometimes, too much good copyright is a bad thing. It's like chilis in soup: a little goes a long way, and too much spoils the broth.

From the Luther Bible to the first phonorecords, from radio to the pulps, from cable to MP3, the world has shown that its first preference for new media is its "democratic-ness" — the ease with which it can reproduced.

(And please, before we get any further, forget all that business about how the Internet's copying model is more disruptive than the technologies that preceded it. For Christ's sake, the Vaudeville performers who sued Marconi for inventing the radio had to go from a regime where they had *one hundred percent* control over who could get into the theater and hear them perform to a regime where they had *zero* percent control over who could build or acquire a radio and tune into a recording of them performing. For that matter, look at the difference between a monkish Bible and a Luther Bible — next to that phase-change, Napster is peanuts.)

Back to democratic-ness. Every successful new medium has traded off its artifact-ness — the degree to which it was populated by bespoke hunks of atoms, cleverly nailed together by master craftspeople — for ease of reproduction. Piano rolls weren't as expressive as good piano players, but they scaled better — as did radio broadcasts, pulp magazines, and MP3s. Liner notes, hand illumination, and leather bindings are nice, but they pale in comparison to the ability of an individual to actually get a copy of her own.

Which isn't to say that old media die. Artists still hand-illumi-nate books; master pianists still stride the boards at Carnegie Hall, and the shelves burst with tell-all biographies of musicians that are richer in detail than any liner-notes booklet. The thing is, when all you've got is monks, every book takes on the character of a monkish Bible. Once you invent the printing press, all the books that are better suited to movable type migrate into that new form. What's left behind are those items that are best suited to the old production scheme: the plays that *need* to be plays, the books that are especially lovely on creamy paper stitched between covers, the music that is most enjoyable performed live and expe-rienced in a throng of humanity.

Increased democratic-ness translates into decreased control: it's a lot harder to control who can copy a book once there's a photocopier on every corner than it is when you need a monas-tery and several years to copy a Bible. And that decreased control demands a new copyright regime that rebalances the rights of creators with their audiences.

For example, when the VCR was invented, the courts affirmed a new copyright exemption for time-shifting; when the radio was invented, the Congress granted an anti-trust exemption to the record labels in order to secure a blanket license; when cable TV was invented, the government just ordered the broadcasters to sell the cable-operators access to programming at a fixed rate.

Copyright is perennially out of date, because its latest rev was generated in response to the last generation of technology. The temptation to treat copyright as though it came down off the mountain on two stone tablets (or worse, as "just like" real prop-erty) is deeply flawed, since, by definition, current copyright only considers the last generation of tech.

So, are bookwarez in violation of copyright law? Duh. Is this the end of the world? *Duh.* If the Catholic church can survive the printing press, science fiction will certainly weather the advent of bookwarez.

## Lagniappe
[Lagniappe]

We're almost done here, but there's one more thing I'd like to do before I get off the stage. [Lagniappe: an unexpected bonus or extra] Think of it as a "lagniappe" — a little something extra to thank you for your patience.

About a year ago, I released my first novel, *Down and Out in the Magic Kingdom,* on the Net, under the terms of the most restrictive Creative Commons license available. All it allowed my readers to do was send around copies of the book. I was cautiously dipping my toe into the water, though at the time, it felt like I was taking a plunge.

Now I'm going to take a plunge. Today, I will re-license the text of *Down and Out in the Magic Kingdom* under a Creative Commons "Attribution-ShareAlike-Derivs-Noncommercial" license [HUMAN READABLE LICENSE], which means that as of today, you have my blessing to create derivative works from my first book. You can make movies, audiobooks, translations, fanfiction, slash fiction (God help us) [GEEK HIERARCHY], furry slash fiction [GEEK HIERARCHY DETAIL], poetry, translations, T-shirts, you name it, with two provisos: that one, you have to allow everyone else to rip, mix, and burn your creations in the same way you're hacking mine; and on the other hand, you've got to do it noncommercially.

The sky didn't fall when I dipped my toe in. Let's see what happens when I get in up to my knees.

The text with the new license will be online before the end of the day. Check craphound.com/down for details.

Oh, and I'm also releasing the text of this speech under a Creative Commons Public Domain dedication, [Public domain dedication] giving it away to the world to do with as it see fits. It'll be linked off my blog, Boing Boing, before the day is through.

# Free(konomic) Ebooks

(Originally published in *Locus*, September 2007.)

Can giving away free electronic books really sell printed books? I think so. As I explained in my March column ("You DO Like Reading Off a Computer Screen"), I don't believe that most readers want to read long-form works off a screen, and I don't believe that they will ever want to read long-form works off a screen. As I say in the column, the problem with reading off a screen isn't resolution, eyestrain, or compatibility with reading in the bathtub: it's that computers are seductive, they tempt us to do other things, making concentrating on a long-form work impractical.

Sure, some readers have the cognitive quirk necessary to read full-length works off screens, or are motivated to do so by other circumstances (such as being so broke that they could never hope to buy the printed work). The rational question isn't, "Will giving away free ebooks cost me sales?" but rather, "Will giving away free ebooks win me more sales than it costs me?"

This is a very hard proposition to evaluate in a quantitative way. Books aren't lattes or cable-knit sweaters: each book sells (or doesn't) due to factors that are unique to that title. It's hard to imagine an empirical, controlled study in which two "equivalent" books are published, and one is also available as a free download, the other not, and the difference calculated as a means of "proving" whether ebooks hurt or help sales in the long run.

I've released all of my novels as free downloads simultaneous with their print publication. If I had a time machine, I could re-release them without the free downloads and compare the royalty statements. Lacking such a device, I'm forced to draw conclusions from qualitative, anecdotal evidence, and I've collected plenty of that:

> Many writers have tried free ebook releases to tie in with the print release of their works. To the best of my knowledge, every writer who's tried this has repeated the experiment with future works, suggesting a high degree of satisfaction with the outcomes.

> A writer friend of mine had his first novel come out at the same time as mine. We write similar material and are often compared to one another by critics and reviewers. My first novel had a free download, his didn't. We compared sales figures and I was doing substantially better than he was — he subsequently convinced his publisher to let him follow suit.

> Baen Books has a pretty good handle on expected sales for new volumes in long-running series; having sold many such series, they have lots of data to use in sales estimates. If Volume N sells X copies, we expect Volume N+1 to sell Y copies. They report that they have seen a measurable uptick in sales following from free ebook releases of previous and current volumes.

> David Blackburn, a Harvard PhD candidate in economics, published a paper in 2004 in which he calculated that, for music, "piracy" results in a net increase in sales for all titles in the 75th percentile and lower; negligible change in sales

for the "middle class" of titles between the 75th percentile
and the 97th percentile; and a small drag on the "super-rich"
in the 97th percentile and higher. Publisher Tim O'Reilly
describes this as "piracy's progressive taxation," apportion-
ing a small wealth-redistribution to the vast majority of
works, no net change to the middle, and a small cost on the
richest few.

> Speaking of Tim O'Reilly, he has just published a detailed,
quantitative study of the effect of free downloads on a
single title. O'Reilly Media published *Asterisk: The Future
of Telephony*, in November 2005, simultaneously releasing
the book as a free download. By March 2007, they had a
pretty detailed picture of the sales-cycle of this book — and,
thanks to industry standard metrics like those provided by
Bookscan, they could compare it, apples-to-apples style,
against the performance of competing books treating with
the same subject. O'Reilly's conclusion: downloads didn't
cause a decline in sales, and appears to have resulted in
a lift in sales. This is particularly noteworthy because the
book in question is a technical reference work, exclusively
consumed by computer programmers who are by definition
disposed to read off screens. Also, this is a reference work
and therefore is more likely to be useful in electronic form,
where it can be easily searched.

> In my case, my publishers have gone back to press repeat-
edly for my books. The print runs for each edition are
modest — I'm a midlist writer in a world with a shrinking
midlist — but publishers print what they think they can sell,
and they're outselling their expectations.

> The new opportunities arising from my free downloads are so numerous as to be uncountable — foreign rights deals, comic book licenses, speaking engagements, article commissions — I've made more money in these secondary markets than I have in royalties.

> More anecdotes: I've had literally thousands of people approach me by email and at signings and cons to say, "I found your work online for free, got hooked, and started buying it." By contrast, I've had all of five emails from people saying, "Hey, idiot, thanks for the free book, now I don't have to buy the print edition, ha ha!"

Many of us have assumed, a priori, that electronic books substitute for print books. While I don't have controlled, quantitative data to refute the proposition, I do have plenty of experience with this stuff, and all that experience leads me to believe that giving away my books is selling the hell out of them.

More importantly, the free ebook skeptics have no evidence to offer in support of their position — just hand-waving and dark muttering about a mythological future when book-lovers give up their printed books for electronic book-readers (as opposed to the much more plausible future where book lovers go on buying their fetish objects *and* carry books around on their electronic devices).

I started giving away ebooks after I witnessed the early days of the "bookwarez" scene, wherein fans cut the binding off their favorite books, scanned them, ran them through optical character recognition software, and manually proofread them to eliminate the digitization errors. These fans were easily spending 80 hours to rip their favorite books, and they were *only* ripping their favorite books, books they loved and wanted to share. (The 80-hour figure comes from my own attempt to do this — I'm sure that rippers get faster with practice.)

I thought to myself that 80 hours' free promotional effort would be a good thing to have at my disposal when my books entered the market. What if I gave my readers clean, canonical electronic editions of my works, saving them the bother of ripping them, and so freed them up to promote my work to their friends?

After all, it's not like there's any conceivable way to stop people from putting books on scanners if they really want to. Scanners aren't going to get more expensive or slower. The Internet isn't going to get harder to use. Better to confront this challenge head on, turn it into an opportunity, than to rail against the future (I'm a science fiction writer — tuning into the future is supposed to be my metier).

The timing couldn't have been better. Just as my first novel was being published, a new, high-tech project for promoting sharing of creative works launched: the Creative Commons project (CC). CC offers a set of tools that make it easy to mark works with whatever freedoms the author wants to give away. CC launched in 2003 and today, more than 160,000,000 works have been released under its licenses.

My next column will go into more detail on what CC is, what licenses it offers, and how to use them — but for now, check them out online at creativecommons.org.

# The Progressive Apocalypse
## and Other Futurismic Delights

(Originally published in *Locus*, July 2007.)

Of course, science fiction is a literature of the present. Many's the science fiction writer who uses the future as a warped mirror for reflecting back the present day, angled to illustrate the hidden strangeness buried by our invisible assumptions: Orwell turned 1948 into *Nineteen Eighty-Four*. But even when the fictional future isn't a parable about the present day, it is necessarily a creation of the present day, since it reflects the present day biases that infuse the author. Hence Asimov's Foundation, a New Deal-esque project to think humanity out of its tribulations through social interventionism.

Bold sf writers eschew the future altogether, embracing a futuristic account of the present day. William Gibson's forthcoming *Spook Country* is an act of "speculative presentism," a book so futuristic it could only have been set in 2006, a book that exploits retrospective historical distance to let us glimpse just how alien and futuristic our present day is.

Science fiction writers aren't the only people in the business of predicting the future. Futurists — consultants, technology columnists, analysts, venture capitalists, and entrepreneurial pitchmen — spill a lot of ink, phosphors, and caffeinated hot air in describing a vision for a future where we'll get more and more of whatever it is they want to sell us or warn us away from. Tomorrow will feature faster, cheaper processors, more Internet users, ubiquitous RFID tags, radically democratic political processes dominated by bloggers, massively multiplayer games whose virtual economies dwarf the physical economy.

There's a lovely neologism to describe these visions: "futurismic." Futurismic media is that which depicts futurism, not the future. It is often self-serving — think of the antigrav Nikes in *Back to the Future III* — and it generally doesn't hold up well to scrutiny.

Sf films and TV are great fonts of futurismic imagery: R2-D2 is a fully conscious AI, can hack the firewall of the Death Star, and is equipped with a range of holographic projectors and antipersonnel devices — but no one has installed a $15 sound card and some text-to-speech software on him, so he has to whistle like Harpo Marx. Or take the Starship *Enterprise*, with a transporter capable of constituting matter from digitally stored plans, and radios that can breach the speed of light.

The non-futurismic version of NCC-1701 would be the size of a softball (or whatever the minimum size for a warp drive, transporter, and subspace radio would be). It would zip around the galaxy at FTL speeds under remote control. When it reached an interesting planet, it would beam a stored copy of a landing party onto the surface, and when their mission was over, it would beam them back into storage, annihilating their physical selves until they reached the next stopping point. If a member of the landing party was eaten by a green-skinned interspatial hippie or giant toga-wearing galactic tyrant, that member would be recovered from backup by the transporter beam. Hell, the entire landing party could consist of multiple copies of the most effective crewmember onboard: no redshirts, just a half-dozen instances of Kirk operating in clonal harmony.

Futurism has a psychological explanation, as recounted in

Harvard clinical psych prof Daniel Gilbert's 2006 book, *Stumbling on Happiness*. Our memories and our projections of the future are necessarily imperfect. Our memories consist of those observations our brains have bothered to keep records of, woven together with inference and whatever else is lying around handy when we try to remember something. Ask someone who's eating a great lunch how breakfast was, and odds are she'll tell you it was delicious. Ask the same question of someone eating rubbery airplane food, and he'll tell you his breakfast was awful. We weave the past out of our imperfect memories and our observable present.

We make the future in much the same way: we use reasoning and evidence to predict what we can, and whenever we bump up against uncertainty, we fill the void with the present day. Hence the injunction on women soldiers in the future of *Starship Troopers,* or the bizarre, glassed-over "Progressland" city diorama at the end of the 1964 World's Fair exhibit "The Carousel of Progress," which Disney built for GE.

Lapsarianism — the idea of a paradise lost, a fall from grace that makes each year worse than the last — is the predominant future feeling for many people. It's easy to see why: an imperfectly remembered golden childhood gives way to the worries of adulthood and physical senescence. Surely the world is getting worse: nothing tastes as good as it did when we were six, everything hurts all the time, and our matured gonads drive us into frenzies of bizarre, self-destructive behavior.

Lapsarianism dominates the Abrahamic faiths. I have an Orthodox Jewish friend whose tradition holds that each generation of rabbis is necessarily less perfect than the rabbis that came before, since each generation is more removed from the perfection of the Garden. Therefore, no rabbi is allowed to overturn any of his forebears' wisdom, since they are all, by definition, smarter than him.

The natural endpoint of Lapsarianism is apocalypse. If things get worse, and worse, and *worse,* eventually they'll just run out of worseness. Eventually, they'll bottom out, a kind of rotten death of the universe when Lapsarian entropy hits the nadir and takes us all with it.

Running counter to Lapsarianism is progressivism: the Enlightenment ideal of a world of great people standing on the shoulders of giants. Each of us contributes to improving the world's storehouse of knowledge (and thus its capacity for bringing joy to all of us), and our descendants and protégés take our work and improve on it. The very idea of "progress" runs counter to the idea of Lapsarianism and the fall: it is the idea that we, as a species, are falling in reverse, combing back the wild tangle of entropy into a neat, tidy braid.

Of course, progress must also have a boundary condition — if only because we eventually run out of imaginary ways that the human condition can improve. And science fiction has a name for the upper bound of progress, a name for the progressive apocalypse:

We call it the Singularity.

Vernor Vinge's Singularity takes place when our technology reaches a stage that allows us to "upload" our minds into software, run them at faster, hotter speeds than our neurological wetware substrate allows for, and create multiple, parallel instances of ourselves. After the Singularity, nothing is predictable because everything is possible. We will cease to be human and become (as the title of Rudy Rucker's next novel would have it) *Postsingular.*

The Singularity is what happens when we have *so much progress* that we run out of progress. It's the apocalypse that ends the human race in rapture and joy. Indeed, Ken MacLeod calls the Singularity "the rapture of the nerds," an apt description for the mirror-world progressive version of the Lapsarian apocalypse.

At the end of the day, both progress and the fall from grace are illusions. The central thesis of *Stumbling on Happiness* is that human beings are remarkably bad at predicting what will make us happy. Our predictions are skewed by our imperfect memories and our capacity for filling the future with the present day.

The future is gnarlier than futurism. NCC-1701 probably wouldn't send out transporter-equipped drones — instead, it would likely find itself on missions whose ethos, mores, and rationale are largely incomprehensible to us, and so obvious to its crew that they couldn't hope to explain them.

Science fiction is the literature of the present, and the present is the only era that we can hope to understand, because it's the only era that lets us check our observations and predictions against reality.

# When the Singularity Is More Than a Literary Device: An Interview with Futurist-Inventor Ray Kurzweil

(Originally published in *Asimov's Science Fiction Magazine,* June 2005.)

It's not clear to me whether the Singularity is a technical belief system or a spiritual one.

The Singularity — a notion that's crept into a lot of skiffy, and whose most articulate in-genre spokesmodel is Vernor Vinge — describes the black hole in history that will be created at the moment when human intelligence can be digitized. When the speed and scope of our cognition is hitched to the price-performance curve of microprocessors, our "progress" will double every eighteen months, and then every twelve months, and then every ten, and eventually, every five seconds.

Singularities are, literally, holes in space from whence no information can emerge, and so sf writers occasionally mutter about how hard it is to tell a story set after the information Singularity. Everything will be different. What it means to be human will be so different that what it means to be in danger, or happy, or sad, or any of the other elements that make up the squeeze-and-release tension in a good yarn will be unrecognizable to us pre-Singletons.

It's a neat conceit to write around. I've committed Singularity a couple of times, usually in collaboration with gonzo Singleton Charlie Stross, the mad antipope of the Singularity. But those stories have the same relation to futurism as romance novels do to love: a shared jumping-off point, but radically different morphologies.

Of course, the Singularity isn't just a conceit for noodling with in the pages of the pulps: it's the subject of serious-minded punditry, futurism, and even science.

Ray Kurzweil is one such pundit-futurist-scientist. He's a serial entrepreneur who founded successful businesses that advanced the fields of optical character recognition (machine-reading) software, text-to-speech synthesis, synthetic musical instrument simulation, computer-based speech recognition, and stock-market analysis. He cured his own Type-II diabetes through a careful review of the literature and the judicious application of first principles and reason. To a casual observer, Kurzweil appears to be the star of some kind of Heinlein novel, stealing fire from the gods and embarking on a quest to bring his maverick ideas to the public despite the dismissals of the establishment, getting rich in the process.

Kurzweil believes in the Singularity. In his 1990 manifesto, "The Age of Intelligent Machines," Kurzweil persuasively argued that we were on the brink of meaningful machine intelligence. A decade later, he continued the argument in a book called *The Age of Spiritual Machines,* whose most audacious claim is that the world's computational capacity has been slowly doubling since the crust first cooled (and before!), and that the doubling interval has been growing shorter and shorter with each passing year, so that now we see it reflected in the computer industry's Moore's Law, which predicts that microprocessors will get twice as powerful for half the cost about every eighteen months. The breathtaking sweep of this trend has an obvious conclusion: computers more powerful than people; more powerful than we can comprehend.

Now Kurzweil has published two more books, *The Singularity Is Near, When Humans Transcend Biology* (Viking, Spring 2005) and *Fantastic Voyage: Live Long Enough to Live Forever* (with Terry Grossman, Rodale, November 2004). The former is a technological roadmap for creating the conditions necessary for ascent into Singularity; the latter is a book about life-prolonging technologies that will assist baby-boomers in living long enough to see the day when technological immortality is achieved.

See what I meant about his being a Heinlein hero?

I still don't know if the Singularity is a spiritual or a technological belief system. It has all the trappings of spirituality, to be sure. If you are pure and kosher, if you live right and if your society is just, then you will live to see a moment of Rapture when your flesh will slough away leaving nothing behind but your ka, your soul, your consciousness, to ascend to an immortal and pure state.

I wrote a novel called *Down and Out in the Magic Kingdom* where characters could make backups of themselves and recover from them if something bad happened, like catching a cold or being assassinated. It raises a lot of existential questions: most prominently: Are you still you when you've been restored from backup?

The traditional AI answer is the Turing Test, invented by Alan Turing, the gay pioneer of cryptography and artificial intelligence who was forced by the British government to take hormone treatments to "cure" him of his homosexuality, culminating in his suicide in 1954. Turing cut through the existentialism about measuring whether a machine is intelligent by proposing a parlor game: a computer sits behind a locked door with a chat program, and a person sits behind another locked door with his own chat program, and they both try to convince a judge that they are real

people. If the computer fools a human judge into thinking that
it's a person, then to all intents and purposes, it's a person.

So how do you know if the backed-up you that you've restored
into a new body — or a jar with a speaker attached to it — is really
you? Well, you can ask it some questions, and if it answers the
same way that you do, you're talking to a faithful copy of yourself.

Sounds good. But the me who sent his first story into *Asimov's*
seventeen years ago couldn't answer the question, "Write a story
for *Asimov's*" the same way the me of today could. Does that mean
I'm not me anymore?

Kurzweil has the answer.

"If you follow that logic, then if you were to take me ten years
ago, I could not pass for myself in a Ray Kurzweil Turing Test.
But once the requisite uploading technology becomes available a
few decades hence, you *could* make a perfect-enough copy of me,
and it *would* pass the Ray Kurzweil Turing Test. The copy doesn't
have to match the quantum state of my every neuron, either: if
you meet me the next day, I'd pass the Ray Kurzweil Turing Test.
Nevertheless, none of the quantum states in my brain would be
the same. There are quite a few changes that each of us undergo
from day to day, we don't examine the assumption that we are the
same person closely.

"We gradually change our pattern of atoms and neurons but
we very rapidly change the particles the pattern is made up of.
We used to think that in the brain — the physical part of us most
closely associated with our identity — cells change very slowly,
but it turns out that the components of the neurons, the tubules,
and so forth, turn over in only *days*. I'm a completely different set
of particles from what I was a week ago.

"Consciousness is a difficult subject, and I'm always surprised
by how many people talk about consciousness routinely as if it

could be easily and readily tested scientifically. But we can't postulate a consciousness detector that does not have some assumptions about consciousness built into it.

"Science is about objective third-party observations and logical deductions from them. Consciousness is about first-person subjective experience, and there's a fundamental gap there. We live in a world of assumptions about consciousness. We share the assumption that other human beings are conscious, for example. But that breaks down when we go outside of humans, when we consider, for example, animals. Some say only humans are conscious and animals are instinctive and machinelike. Others see humanlike behavior in an animal and consider the animal conscious, but even these observers don't generally attribute consciousness to animals that aren't humanlike.

"When machines are complex enough to have responses recognizable as emotions, those machines will be more humanlike than animals."

The Kurzweil Singularity goes like this: computers get better and smaller. Our ability to measure the world gains precision and grows ever cheaper. Eventually, we can measure the world inside the brain and make a copy of it in a computer that's as fast and complex as a brain, and voila, intelligence.

Here in the twenty-first century we like to view ourselves as ambulatory brains, plugged into meat-puppets that lug our precious gray matter from place to place. We tend to think of that gray matter as transcendently complex, and we think of it as being the bit that makes us *us*.

But brains aren't that complex, Kurzweil says. Already, we're starting to unravel their mysteries.

"We seem to have found one area of the brain closely associated with higher-level emotions, the spindle cells, deeply embedded in the brain. There are tens of thousands of them, spanning the whole brain (maybe eighty thousand in total), which is an incredibly small number. Babies don't have any, most animals don't have any, and they likely only evolved over the last million years or so. Some of the high-level emotions that are deeply human come from these.

"Turing had the right insight: base the test for intelligence on written language. Turing Tests really work. A novel is based on language: with language you can conjure up any reality, much more so than with images. Turing almost lived to see computers doing a good job of performing in fields like math, medical diagnosis, and so on, but those tasks were easier for a machine than demonstrating even a child's mastery of language. Language is the true embodiment of human intelligence."

If we're not so complex, then it's only a matter of time until computers are more complex than us. When that comes, our brains will be model-able in a computer and that's when the fun begins. That's the thesis of *Spiritual Machines,* which even includes a (Heinlein-style) timeline leading up to this day.

Now, it may be that a human brain contains $n$ logic-gates and runs at $x$ cycles per second and stores $z$ petabytes, and that $n$ and $x$ and $z$ are all within reach. It may be that we can take a brain apart and record the position and relationships of all the neurons and sub-neuronal elements that constitute a brain.

But there are also a nearly infinite number of ways of modeling a brain in a computer, and only a finite (or possibly nonexistent) fraction of that space will yield a conscious copy of the original meat-brain. Science fiction writers usually hand-wave this step: in Heinlein's "The Man Who Sold the Moon," the gimmick is that once the computer becomes complex enough, with enough "random numbers," it just wakes up.

Computer programmers are a little more skeptical. Computers have never been known for their skill at programming themselves — they tend to be no smarter than the people who write their software.

But there are techniques for getting computers to program themselves, based on evolution and natural selection. A programmer creates a system that spits out lots — thousands or even millions — of randomly generated programs. Each one is given the opportunity to perform a computational task (say, sorting a list of numbers from greatest to least) and the ones that solve the problem best are kept aside while the others are erased. Now the survivors are used as the basis for a new generation of randomly mutated descendants, each based on elements of the code that preceded them. By running many instances of a randomly varied program at once, and by culling the least successful and regenerating the population from the winners very quickly, it is possible to *evolve* effective software that performs as well or better than the code written by human authors.

Indeed, evolutionary computing is a promising and exciting field that's realizing real returns through cool offshoots like "ant colony optimization" and similar approaches that are showing good results in fields as diverse as piloting military UAVs and efficiently provisioning car-painting robots at automotive plants.

So if you buy Kurzweil's premise that computation is getting cheaper and more plentiful than ever, then why not just use evolutionary algorithms to *evolve* the best way to model a scanned-in human brain such that it "wakes up" like Heinlein's Mike computer?

Indeed, this is the crux of Kurzweil's argument in *Spiritual Machines*: if we have computation to spare and a detailed model of a human brain, we need only combine them and out will pop the mechanism whereby we may upload our consciousness to digital storage media and transcend our weak and bothersome meat forever.

But it's a cheat. Evolutionary algorithms depend on the same mechanisms as real-world evolution: heritable variation of candidates and a system that culls the least-suitable candidates. This latter — the fitness-factor that determines which individuals in a cohort breed and which vanish — is the key to a successful evolutionary system. Without it, there's no pressure for the system to achieve the desired goal: merely mutation and more mutation.

But how can a machine evaluate which of a trillion models of a human brain is "most like" a conscious mind? Or better still: Which one is most like the individual whose brain is being modeled?

"It is a sleight of hand in *Spiritual Machines*," Kurzweil admits. "But in *The Singularity Is Near,* I have an in-depth discussion about what we know about the brain and how to model it. Our tools for understanding the brain are subject to the Law of Accelerating Returns, and we've made more progress in reverse-engineering the human brain than most people realize." This is a tasty Kurzweilism that observes that improvements in technology yield tools for improving technology, round and round, so that the thing that progress begets more than anything is more and yet faster progress.

"Scanning resolution of human tissue — both spatial and temporal — is doubling every year, and so is our knowledge of the workings of the brain. The brain is not one big neural net, the brain is several hundred different regions, and we can understand each region, we can model the regions with mathematics, most of which have some nexus with chaos and self-organizing systems. This has already been done for a couple dozen regions out of the several hundred.

"We have a good model of a dozen or so regions of the auditory and visual cortex, how we strip images down to very low-resolution movies based on pattern recognition. Interestingly, we don't actually see things, we essentially hallucinate them in detail from what we see from these low resolution cues. Past the early phases of the visual cortex, detail doesn't reach the brain.

"We are getting *exponentially* more knowledge. We can get detailed scans of neurons working *in vivo,* and are beginning to understand the chaotic algorithms underlying human intelligence. In some cases, we are getting comparable performance of brain regions in simulation. These tools will continue to grow in detail and sophistication.

"We can have confidence of reverse-engineering the brain in twenty years or so. The reason that brain reverse-engineering has not contributed much to artificial intelligence is that up until recently we didn't have the right tools. If I gave you a computer and a few magnetic sensors and asked you to reverse-engineer it, you might figure out that there's a magnetic device spinning when a file is saved, but you'd never get at the instruction set. Once you reverse-engineer the computer fully, however, you can express its principles of operation in just a few dozen pages.

"Now there are new tools that let us see the interneuronal connections and their signaling, *in vivo,* and in real-time. We're just now getting these tools and there's very rapid application of the tools to obtain the data.

"Twenty years from now we will have realistic simulations and models of all the regions of the brain and [we will] understand how they work. We won't blindly or mindlessly copy those methods, we will understand them and use them to improve our AI toolkit. So we'll learn how the brain works and then apply the sophisticated tools that we will obtain, as we discover how the brain works.

"Once we understand a subtle science principle, we can isolate, amplify, and expand it. Air goes faster over a curved surface: from that insight we isolated, amplified, and expanded the idea and invented air travel. We'll do the same with intelligence.

"Progress is exponential — not just a measure of power of computation, number of Internet nodes, and magnetic spots on a hard disk — the rate of paradigm shift is itself accelerating, doubling every decade. Scientists look at a problem and they intuitively conclude that since we've solved 1 percent over the last year, it'll therefore be one hundred years until the problem is exhausted: but the rate of progress doubles every decade, and the power of the information tools (in price-performance, resolution, bandwidth, and so on) doubles every year. People, even scientists, don't grasp exponential growth. During the first decade of the human genome project, we only solved 2 percent of the problem, but we solved the remaining 98 percent in five years."

But Kurzweil doesn't think that the future will arrive in a rush. As William Gibson observed, "The future is here, it's just not evenly distributed."

"Sure, it'd be interesting to take a human brain, scan it, reinstantiate the brain, and run it on another substrate. That will ultimately happen.

"But the most salient scenario is that we'll *gradually* merge with our technology. We'll use nanobots to kill pathogens, then to kill cancer cells, and then they'll go into our brain and do benign things there like augment our memory, and very gradually they'll get more and more sophisticated. There's no single great leap, but there is ultimately a great leap comprised of many small steps.

"In *The Singularity Is Near*, I describe the radically different world of 2040, and how we'll get there one benign change at a time. The Singularity will be gradual, smooth.

"Really, this is about augmenting our biological thinking with nonbiological thinking. We have a capacity of 1026 to 1029 calculations per second (cps) in the approximately 1010 biological human brains on Earth and that number won't change much in fifty years, but nonbiological thinking will just crash through that. By 2049, nonbiological thinking capacity will be on the order of a billion times that. We'll get to the point where bio thinking is relatively insignificant.

"People didn't throw their typewriters away when word-processing started. There's always an overlap — it'll take time before we realize how much more powerful nonbiological thinking will ultimately be."

It's well and good to talk about all the stuff we *can* do with technology, but it's a lot more important to talk about the stuff we'll be *allowed* to do with technology. Think of the global freak-out caused by the relatively trivial advent of peer-to-peer file-sharing tools: universities are wiretapping their campuses and disciplining computer science students for writing legitimate, general purpose software; grandmothers and twelve-year-olds are losing their life savings; privacy and due process have sailed out the window without so much as a by-your-leave.

Even P2P's worst enemies admit that this is a general-purpose technology with good *and* bad uses, but when new tech comes along it often engenders a response that countenances punishing an infinite number of innocent people to get at the guilty.

What's going to happen when the new technology paradigm isn't song-swapping, but transcendent super-intelligence? Will the reactionary forces be justified in razing the whole ecosystem to eliminate a few parasites who are doing negative things with the new tools?

"Complex ecosystems will always have parasites. Malware [malicious software] is the most important battlefield today.

"*Everything* will become software — objects will be malleable, we'll spend lots of time in VR, and computhought will be orders of magnitude more important than biothought.

"Software is already complex enough that we have an ecological terrain that has emerged just as it did in the bioworld.

"That's partly because technology is unregulated and people have access to the tools to create malware and the medicine to treat it. Today's software viruses are clever and stealthy and not simpleminded. *Very* clever.

"But here's the thing: you don't see people advocating shutting down the Internet because malware is so destructive. I mean, malware is potentially more than a nuisance — emergency systems, air traffic control, and nuclear reactors all run on vulnerable software. It's an important issue, but the potential damage is still a tiny fraction of the benefit we get from the Internet.

"I hope it'll remain that way — that the Internet won't become a regulated space like medicine. Malware's not the most important issue facing human society today. Designer bioviruses are. People are concerned about WMDs, but the most daunting WMD would be a designed biological virus. The means exist in college labs to create destructive viruses that erupt and spread silently with long incubation periods.

"Importantly, a would-be bio-terrorist doesn't have to put malware through the FDA's regulatory approval process, but scientists working to fix bio-malware *do*.

"In Huxley's *Brave New World*, the rationale for the totalitarian system was that technology was too dangerous and needed to be controlled. But that just pushes technology underground where it becomes *less* stable. Regulation gives the edge of power to the irresponsible who won't listen to the regulators anyway.

"The way to put more stones on the defense side of the scale is to put more resources into defensive technologies, not create a totalitarian regime of Draconian control.

"I advocate a one hundred billion dollar program to accelerate the development of anti-biological virus technology. The way to combat this is to develop broad tools to destroy viruses. We have tools like RNA interference, just discovered in the past two years to block gene expression. We could develop means to sequence the genes of a new virus (SARS only took thirty-one days) and respond to it in a matter of days.

"Think about it. There's no FDA for software, no certification for programmers. The government is thinking about it, though! The reason the FCC is contemplating Trusted Computing mandates" — a system to restrict what a computer can do by means of hardware locks embedded on the motherboard — "is that computing technology is broadening to cover everything. So now you have communications bureaucrats, biology bureaucrats, all wanting to regulate computers.

"Biology would be a lot more stable if we moved away from regulation — which is extremely irrational and onerous and doesn't appropriately balance risks. Many medications are not available today even though they should be. The FDA always wants to know what happens if we approve this and will it turn into a thalidomide situation that embarrasses us on CNN?

"Nobody asks about the harm that will certainly accrue from delaying a treatment for one or more years. There's no political weight at all, people have been dying from diseases like heart disease and cancer for as long as we've been alive. Attributable risks get 100–1000 times more weight than unattributable risks."

Is this spirituality or science? Perhaps it is the melding of both — more shades of Heinlein, this time the weird religions founded by people who took *Stranger in a Strange Land* way too seriously.

After all, this is a system of belief that dictates a means by which we can care for our bodies virtuously and live long enough to transcend them. It is a system of belief that concerns itself with the meddling of non-believers, who work to undermine its goals through irrational systems predicated on their disbelief. It is a system of belief that asks and answers the question of what it means to be human.

It's no wonder that the Singularity has come to occupy so much of the science fiction narrative in these years. Science or spirituality, you could hardly ask for a subject better tailored to technological speculation and drama.

# Wikipedia: A Genuine H2G2 — Minus the Editors

(Originally published in *The Anthology at the End of the Universe: Leading Science Fiction Authors on Douglas Adams' The Hitchhiker's Guide to the Galaxy*, edited by Glenn Yeffeth and Shauna Caughey.)

"Mostly Harmless" — a phrase so funny that Adams actually titled a book after it. Not that there's a lot of comedy inherent in those two words: rather, they're the punchline to a joke that anyone who's ever written for publication can really get behind.

Ford Prefect, a researcher for *The Hitchhiker's Guide to the Galaxy*, has been stationed on Earth for years, painstakingly compiling an authoritative, insightful entry on Terran geography, science and culture, excerpts from which appear throughout the *H2G2* books. His entry improved upon the old one, which noted that Earth was, simply, "Harmless."

However, the *Guide* has limited space, and when Ford submits his entry to his editors, it is trimmed to fit:

"What? Harmless? Is that all it's got to say? Harmless! One word!"

Ford shrugged. "Well, there are a hundred billion stars in the Galaxy, and only a limited amount of space in the book's microprocessors," he said, "and no one knew much about the Earth of course."

"Well for God's sake I hope you managed to rectify that a bit."

"Oh yes, well I managed to transmit a new entry off to the editor. He had to trim it a bit, but it's still an improvement."

"And what does it say now?" asked Arthur.

"Mostly harmless," admitted Ford with a slightly embarrassed cough.

[fn: My lifestyle is as gypsy and fancy-free as the characters in *H2G2*, and as a result my copies of the Adams books are thousands of miles away in storages in other countries, and this essay was penned on public transit and in cheap hotel rooms in Chile, Boston, London, Geneva, Brussels, Bergen, Geneva (again), Toronto, Edinburgh, and Helsinki. Luckily, I was able to download a dodgy, re-keyed version of the Adams books from a peer-to-peer network, which I accessed via an open wireless network on a random street-corner in an anonymous city, a fact that I note here as testimony to the power of the Internet to do what the *Guide* does for Ford *and* Arthur: put all the information I need at my fingertips, wherever I am. However, these texts are a little on the dodgy side, as noted, so you might want to confirm these quotes before, say, uttering them before an Adams truefan.]

And there's the humor: every writer knows the pain of laboring over a piece for days, infusing it with diverse interesting factoids and insights, only to have it cut to ribbons by some distant editor. (I once wrote thirty drafts of a 5,000-word article for an editor who ended up running it in three paragraphs as accompaniment for what he decided should be a photo essay with minimal verbiage.)

Since the dawn of the Internet, *H2G2* geeks have taken it upon themselves to attempt to make a *Guide* on the Internet. Volunteers wrote and submitted essays on various subjects as would be likely to appear in a good encyclopedia, infusing them with equal

measures of humor and thoughtfulness, and they were edited together by the collective effort of the contributors. These projects
— Everything2, H2G2 (which was overseen by Adams himself),
and others — are like a barn-raising in which a team of dedicated
volunteers organize the labors of casual contributors, piecing together a free and open user-generated encyclopedia.

These encyclopedias have one up on Adams's *Guide:* they have
no shortage of space on their "microprocessors" (the first volume
of the *Guide* was clearly written before Adams became conversant
with PCs!). The ability of humans to generate verbiage is far out-
stripped by the ability of technologists to generate low-cost, reliable storage to contain it. For example, Brewster Kahle's Internet
Archive project (archive.org) has been making a copy of the Web
— the *whole* Web, give or take — every couple of days since 1996.
Using the Archive's Wayback Machine, you can now go and see
what any page looked like on a given day.

The Archive doesn't even bother throwing away copies of pages
that haven't changed since the last time they were scraped: with
storage as cheap as it is — and it is *very* cheap for the Archive,
which runs the largest database in the history of the universe off
of a collection of white-box commodity PCs stacked up on pack-
ing skids in the basement of a disused armory in San Francisco's
Presidio — there's no reason not to just keep them around. In fact,
the Archive has just spawned two "mirror" Archives, one located
under the rebuilt Library of Alexandria and the other in Amster-
dam. [fn: Brewster Kahle says that he was nervous about keeping his only copy of the "repository of all human knowledge" on
the San Andreas fault, but keeping your backups in a censorship-
happy Amnesty International watchlist state and/or in a flood-
plain below sea level is probably not such a good idea either!]

So these systems did not see articles trimmed for lack of space;
for on the Internet, the idea of "running out of space" is meaning-

less. But they *were* trimmed, by editorial cliques, and rewritten for clarity and style. Some entries were rejected as being too thin, while others were sent back to the author for extensive rewrites.

This traditional separation of editor and writer mirrors the creative process itself, in which authors are exhorted to concentrate on *either* composing *or* revising, but not both at the same time, for the application of the critical mind to the creative process strangles it. So you write, and then you edit. Even when you write for your own consumption, it seems you have to answer to an editor.

The early experimental days of the Internet saw much experimentation with alternatives to traditional editor/author divisions. Slashdot, a nerdy news-site of surpassing popularity [fn: Having a link to one's website posted to Slashdot will almost inevitably overwhelm your server with traffic, knocking all but the best-provisioned hosts offline within minutes; this is commonly referred to as "the Slashdot Effect."], has a baroque system for "community moderation" of the responses to the articles that are posted to its front pages. Readers, chosen at random, are given five "moderator points" that they can use to raise or lower the score of posts on the Slashdot message-boards. Subsequent readers can filter their views of these boards to show only highly ranked posts. Other readers are randomly presented with posts and their rankings and are asked to rate the fairness of each moderator's moderation. Moderators who moderate fairly are given more opportunities to moderate; likewise message-board posters whose messages are consistently highly rated.

It is thought that this system rewards good "citizenship" on the Slashdot boards through checks and balances that reward good messages and fair editorial practices. And in the main, the Slashdot moderation system works [fn: as do variants on it, like the system in place at Kur5hin.org (pronounced "corrosion")]. If you dial your filter up to show you highly scored messages, you will generally get well-reasoned, or funny, or genuinely useful posts in your browser.

This community moderation scheme and ones like it have been heralded as a good alternative to traditional editorship. The importance of the Internet to "edit itself" is best understood in relation to the old shibboleth, "On the Internet, everyone is a slushreader." [fn: "Slush" is the term for generally execrable unsolicited manuscripts that fetch up in publishers' offices — these are typically so bad that the most junior people on staff are drafted into reading (and, usually, rejecting) them.] When the Internet's radical transformative properties were first bandied about in publishing circles, many reassured themselves that even if printing's importance was de-emphasized, that good editors would always be needed, and doubly so online, where any mouth-breather with a modem could publish his words. Someone would need to separate the wheat from the chaff and help keep us from drowning in information.

One of the best-capitalized businesses in the history of the world, Yahoo!, went public on the strength of this notion, proposing to use an army of researchers to catalog every single page on the Web even as it was created, serving as a comprehensive guide to all human knowledge. Less than a decade later, Yahoo! is all but out of that business: the ability of the human race to generate new pages far outstrips Yahoo!'s ability to read, review, rank, and categorize them.

Hence Slashdot, a system of distributed slushreading. Rather than professionalizing the editorship role, Slashdot invites contributors to identify good stuff when they see it, turning editorship into a reward for good behavior.

But as well as Slashdot works, it has this signal failing: nearly every conversation that takes place on Slashdot is shot through with discussion, griping, and gaming *on the moderation system itself*. The core task of Slashdot has *become* editorship, not the putative subjects of Slashdot posts. The fact that the central task of Slashdot is to rate other Slashdotters creates a tenor of meanness in the discussion. Imagine if the subtext of every discussion you had in the real world was a kind of running, pedantic nitpickery in which every point was explicitly weighed and judged and commented upon. You'd be an unpleasant, unlikable jerk, the kind of person that is sometimes referred to as a "slashdork."

As radical as Yahoo!'s conceit was, Slashdot's was more radical. But as radical as Slashdot's is, it is still inherently conservative in that it presumes that editorship is necessary, and that it further requires human judgment and intervention.

Google's a lot more radical. Instead of editors, it has an algorithm. Not the kind of algorithm that dominated the early search engines like AltaVista, in which laughably bad artificial intelligence engines attempted to automatically understand the content, context, and value of every page on the Web so that a search for "Dog" would turn up the page more relevant to the query.

Google's algorithm is predicated on the idea that people are good at understanding things and computers are good at counting things. Google counts up all the links on the Web and affords more authority to those pages that have been linked to by the most other pages. The rationale is that if a page has been linked to by many web-authors, then they must have seen some merit in that page. This system works remarkably well — so well that it's nearly inconceivable that any search-engine would order its rankings by any other means. What's more, it doesn't pervert the tenor of the discussions and pages that it catalogs by turning each one into a performance for a group of ranking peers. [fn: Or at least, it *didn't*. Today, dedicated web-writers, such as bloggers, are keenly aware of the way that Google will interpret their choices about linking and page-structure. One popular sport is "googlebombing," in which web-writers collude to link to a given page using a humorous keyword so that the page becomes the top result for that word — which is why, for a time, the top result for "more evil than Satan" was Microsoft.com. Likewise, the practice of "blogspamming," in which unscrupulous spammers post links to their webpages in the message-boards on various blogs, so that Google will be tricked into thinking that a wide variety of sites have conferred some authority onto their penis-enlargement page.]

But even Google is conservative in assuming that there is a need for editorship as distinct from composition. Is there a way we can dispense with editorship altogether and just use composition to refine our ideas? Can we merge composition and editorship into a single role, fusing our creative and critical selves?

You betcha.

"Wikis" [fn: Hawai'ian for "fast"] are websites that can be edited by anyone. They were invented by Ward Cunningham in 1995, and they have become one of the dominant tools for Internet collaboration in the present day. Indeed, there is a sort of Internet geek who throws up a Wiki in the same way that ants make anthills: reflexively, unconsciously.

Here's how a Wiki works. You put up a page:

> Welcome to my Wiki. It is rad.
> There are OtherWikis that inspired me.

Click "publish" and bam, the page is live. The word "OtherWikis" will be underlined, having automatically been turned into a link to a blank page titled "OtherWikis." (Wiki software recognizes words with capital letters in the middle of them as links to other pages. Wiki people call this "camel-case," because the capital letters in the middle of words make them look like humped camels.) At the bottom of it appears this legend: "Edit this page."

Click on "Edit this page" and the text appears in an editable field. Revise the text to your heart's content and click "Publish" and your revisions are live. Anyone who visits a Wiki can edit any of its pages, adding to it, improving on it, adding camel-cased links to new subjects, or even defacing or deleting it.

It is authorship without editorship. Or authorship fused with editorship. Whichever, it works, though it requires effort. The Internet, like all human places and things, is fraught with spoilers and vandals who deface whatever they can. Wiki pages are routinely replaced with obscenities, with links to spammers' websites, with junk and crap and flames.

But Wikis have self-defense mechanisms, too. Anyone can "subscribe" to a Wiki page, and be notified when it is updated. Those who create Wiki pages generally opt to act as "gardeners" for them, ensuring that they are on hand to undo the work of the spoilers.

In this labor, they are aided by another useful Wiki feature: the "history" link. Every change to every Wiki page is logged and recorded. Anyone can page back through every revision, and anyone can revert the current version to a previous one. That means that vandalism only lasts as long as it takes for a gardener to come by and, with one or two clicks, set things to right.

This is a powerful and wildly successful model for collaboration, and there is no better example of this than the Wikipedia, a free, Wiki-based encyclopedia with more than one million entries, which has been translated into 198 languages [fn: That is, one or more Wikipedia entries have been translated into 198 languages; more than 15 languages have 10,000 or more entries translated.]

Wikipedia is built entirely out of Wiki pages created by self-appointed experts. Contributors research and write up subjects, or produce articles on subjects that they are familiar with.

This is authorship, but what of editorship? For if there is one thing a *Guide* or an encyclopedia must have, it is authority. It must be vetted by trustworthy, neutral parties, who present something that is either The Truth or simply A Truth, but truth nevertheless.

The Wikipedia has its skeptics. Al Fasoldt, a writer for the *Syracuse Post-Standard,* apologized to his readers for having recommended that they consult Wikipedia. A reader of his, a librarian, wrote in and told him that his recommendation had been irresponsible, for Wikipedia articles are often defaced, or worse still, rewritten with incorrect information. When another journalist from the Techdirt website wrote to Fasoldt to correct this impression, Fasoldt responded with an increasingly patronizing and hysterical series of messages in which he described Wikipedia as "outrageous," "repugnant," and "dangerous," insulting the Techdirt writer and storming off in a huff. [fn: see http://techdirt.com/articles/20040827/0132238_F.shtml for more]

Spurred on by this exchange, many of Wikipedia's supporters decided to empirically investigate the accuracy and resilience of the system. Alex Halavais made changes to thirteen different pages, ranging from obvious to subtle. Every single change was found and corrected within hours. [fn: see http://alex.halavais.net/news/index.php?p=794 for more] Then legendary Princeton engineer Ed Felten ran side-by-side comparisons of Wikipedia entries on areas in which he had deep expertise with their counterparts in the current electronic edition of the *Encyclopedia Britannica.* His conclusion? "Wikipedia's advantage is in having more, longer, and more current entries. If it weren't for the Microsoft-case entry, Wikipedia would have been the winner hands down. *Britannica*'s advantage is in having lower variance in the quality of its entries." [fn: see http://www.freedom-to-tinker.com/ar-chives/000675.html for more]  Not a complete win for Wikipedia, but hardly "outrageous," "repugnant," and "dangerous." (Poor Fasoldt — his idiotic hyperbole will surely haunt him through the whole of his career — I mean, "repugnant"?!)

There has been one very damning and even frightening indictment of Wikipedia, which came from Ethan Zuckerman, the founder of the Geekcorps group, which sends volunteers to poor countries to help establish Internet Service Providers and do other good works through technology.

Zuckerman, a Harvard Berkman Center Fellow, is concerned with the "systemic bias" in a collaborative encyclopedia whose contributors must be conversant with technology and in possession of same in order to improve on the work there. Zuckerman reasonably observes that Internet users skew towards wealth, residence in the world's richest countries, and a technological bent. This means that the Wikipedia, too, is skewed to subjects of interest to that group — subjects where that group already has expertise and interest.

The result is tragicomical. The entry on the Congo Civil War, the largest military conflict the world has seen since WWII, which has claimed over three million lives, has only a fraction of the verbiage devoted to the War of the Ents, a fictional war fought between sentient trees in J. R. R. Tolkien's *The Lord of the Rings*.

Zuckerman issued a public call to arms to rectify this, challenging Wikipedia contributors to seek out information on subjects like Africa's military conflicts, nursing, and agriculture and write these subjects up in the same loving detail given over to science fiction novels and contemporary youth culture. His call has been answered well. What remains is to infiltrate the Wikipedia into the academe so that term papers, Masters and Doctoral theses on these subjects find themselves in whole or in part on the Wikipedia. [fn See *http://en.wikipedia.org/wiki/User:Xed/CROSSBOW* for more on this]

But if Wikipedia is authoritative, how does it get there? What alchemy turns the maunderings of "mouth-breathers with modems" into valid, useful encyclopedia entries?

It all comes down to the way that disputes are deliberated over and resolved. Take the entry on Israel. At one point, it characterized Israel as a beleaguered state set upon by terrorists who would drive its citizens into the sea. Not long after, the entry was deleted holus-bolus and replaced with one that described Israel as an illegal state practicing Apartheid on an oppressed ethnic minority.

Back and forth the editors went, each overwriting the other's with his or her own doctrine. But eventually, one of them blinked. An editor moderated the doctrine just a little, conceding a single point to the other. And the other responded in kind. In this way, turn by turn, all those with a strong opinion on the matter negotiated a kind of Truth, a collection of statements that everyone could agree represented as neutral a depiction of Israel as was likely to emerge. Whereupon, the joint authors of this marvelous document joined forces and fought back-to-back to resist the revisions of other doctrinaires who came later, preserving their hard-won peace. [fn: This process was just repeated in microcosm in the Wikipedia entry on the author of this paper, which was replaced by a rather disparaging and untrue entry that characterized his books as critical and commercial failures — there ensued several editorial volleys, culminating in an uneasy peace that couches the anonymous detractor's skepticism in context and qualifiers that make it clear what the facts are and what is speculation.]

What's most fascinating about these entries isn't their "final" text as currently present on Wikipedia. It is the history page for each, blow-by-blow revision lists that make it utterly transparent where the bodies were buried on the way to arriving at whatever Truth has emerged. This is a neat solution to the problem of authority — if you want to know what the fully rounded view of opinions on any controversial subject look like, you need only consult its entry's history page for a blistering eyeful of thorough debate on the subject.

And here, finally, is the answer to the "Mostly harmless" problem. Ford's editor can trim his verbiage to two words, but they need not stay there — Arthur, or any other user of the *Guide* as we know it today [fn: that is, in the era where we understand enough about technology to know the difference between a microprocessor and a hard drive] can revert to Ford's glorious and exhaustive version.

Think of it: a *Guide* without space restrictions and without editors, where any Vogon can publish to his heart's content.

Lovely.

# Warhol is Turning in His Grave

(Originally published in *The Guardian,* November 13, 2007.)

The excellent program for Pop Art Portraits, the current exhibition at London's National Portrait Gallery, has a lot to say about the pictures hanging on the walls and the diverse source material the artists used to produce their provocative works.

Apparently they cut up magazines, copied comic books, drew trademarked cartoon characters like Minnie Mouse, reproduced covers from *Time* magazine, made ironic use of a cartoon Charles Atlas, painted over iconic photos of James Dean and Elvis Presley — and that's just in the first of seven rooms.

The program describes the aesthetic experience conjured up by these transmogrified icons of high and low culture. Celebrated pop artists including Larry Poons, Robert Rauschenberg, and Andy Warhol created these images by nicking the work of others, without permission, and transforming it to make statements and evoke emotions never countenanced by the original creators.

Despite this, the program does not say a word about copyright. Can you blame the authors? A treatise on the way that copyright and trademarks were — *had to be* — trammeled to make these works could fill volumes.

Reading the program, you can only assume that the curators' message about copyright is that where free expression is concerned, the rights of the creators of the original source material must take a back seat to those of the pop artists.

There is, however, another message about copyright in the

National Portrait Gallery: it is implicit in the "No Photography" signs prominently displayed throughout its rooms, including one by the entrance to the Pop Art Portraits exhibition.

These signs are not intended to protect the works from the depredations of camera flashes (otherwise they would read "No Flash Photography"). No, the ban on pictures is meant to safeguard the copyright of the works hung on the walls — a fact that every member of staff I asked instantly confirmed.

Indeed, it seems every square centimeter of the National Portrait Gallery is under some form of copyright. I wasn't even allowed to photograph the "No Photography" sign. A member of staff explained that the typography and layout of the signs was itself copyrighted.

If true, presumably the same rules would prevent anyone from taking any pictures in any public place — unless you could somehow contrive to get a shot of Leicester Square without any writing, logos, architectural facades, or images in it. Otherwise I doubt even Warhol could have gotten away with it.

So what's the message of the show? Is it a celebration of remix culture, reveling in the endless possibilities opened up by appropriating and reusing images without permission?

Or is it the epitaph on the tombstone of the sweet days before the UN set up the World Intellectual Property Organization and the ensuing mania for turning everything that can be sensed and recorded into someone's property?

Does this show — paid for with public money, with some works that are themselves owned by public institutions — seek to inspire us to become twenty-first century pop artists, armed with cameraphones, websites, and mixers, or is it supposed to inform us that our chance has passed and we'd best settle for a life as information serfs who can't even make free use of what our eyes see and our ears hear?

Perhaps, just perhaps, this is actually a Dadaist show *masquerading* as a pop art show. Perhaps the point is to titillate us with the delicious irony of celebrating copyright infringement while simultaneously taking the view that even the "No Photography" sign is a form of property not to be reproduced without the permission that can never be had.

# The Future of Ignoring Things

(Originally published on *InformationWeek*'s Internet Evolution [*www.
internetevolution.com*], October 3, 2007.)

For decades, computers have been helping us to remember, but
now it's time for them to help us to ignore.

Take email: Endless engineer-hours are poured into stopping
spam, but virtually no attention is paid to our interaction with
our non-spam messages. Our mailer may strive to learn from
our ratings what is and is not spam, but it expends practically no
effort on figuring out which of the non-spam emails are important and which ones can be safely ignored, dropped into archival
folders, or deleted unread.

For example, I'm forever getting cc'd on busy threads by well-
meaning colleagues who want to loop me in on some discussion
in which I have little interest. Maybe the initial group invitation
to a dinner (that I'll be out of town for) was something I needed
to see, but now that I've declined, I really don't need to read the
300+ messages that follow debating the best place to eat.

I could write a mail-rule to ignore the thread, of course. But
mail-rule editors are clunky, and once your rule-list grows very
long, it becomes increasingly unmanageable. Mail-rules are where
bookmarks were before the bookmark site del.icio.us showed up
— built for people who might want to ensure that messages from
the boss show up in red, but not intended to be used as a gigantic
storehouse of a million filters, a crude means for telling the computers what we don't want to see.

Rael Dornfest, the former chairman of the O'Reilly Emerging Technology Conference and founder of the startup IWantSandy, once proposed an "ignore thread" feature for mailers: Flag a thread as uninteresting, and your mailer will start to hide messages with that subject-line or thread-ID for a week, unless those messages contain your name. The problem is that threads mutate. Last week's dinner plans become this week's discussion of next year's group holiday. If the thread is still going after a week, the messages flow back into your inbox — and a single click takes you back through all the messages you missed.

We need a million measures like this, adaptive systems that create a gray zone between "delete on sight" and "show this to me right away."

RSS readers are a great way to keep up with the torrent of new items posted on high-turnover sites like Digg, but they're even better at keeping up with sites that are sporadic, like your friend's brilliant journal that she only updates twice a year. But RSS readers don't distinguish between the rare and miraculous appearance of a new item in an occasional journal and the latest click-fodder from Slashdot. They don't even sort your RSS feeds according to the sites that you click-through the most.

There was a time when I could read the whole of Usenet — not just because I was a student looking for an excuse to avoid my assignments, but because Usenet was once tractable, readable by a single determined person. Today, I can't even keep up with a single high-traffic message-board. I can't read all my email. I can't read every item posted to every site I like. I certainly can't plough through the entire edit-history of every Wikipedia entry I read. I've come to grips with this — with acquiring information on a probabilistic basis, instead of the old, deterministic, cover-to-cover approach I learned in the offline world.

It's as though there's a cognitive style built into TCP/IP. Just as the network only does best-effort delivery of packets, not worry-ing so much about the bits that fall on the floor, TCP/IP users also do best-effort sweeps of the Internet, focusing on learning from the good stuff they find, rather than lamenting the stuff they don't have time to see.

The network won't ever become more tractable. There will never be fewer things vying for our online attention. The only answer is better ways and new technology to ignore stuff — a field that's just being born, with plenty of room to grow.

# Facebook's Faceplant

(Originally published as "How Your Creepy Ex-Co-Workers Will Kill Facebook," *InformationWeek,* November 26, 2007.)

Facebook's "platform" strategy has sparked much online debate and controversy. No one wants to see a return to the miserable days of walled gardens, when you couldn't send a message to an AOL subscriber unless you, too, were a subscriber, and when the only services that made it were the ones that AOL management approved. Those of us on the "real" Internet regarded AOL with a species of superstitious dread, a hive of clueless noobs waiting to swamp our beloved Usenet with dumb flamewars (we fiercely guarded our erudite flamewars as being of a palpably superior grade), the wellspring of an endless geyser of free floppy disks and CDs, the kind of place where the clueless management were willing and able to — for example — alienate every Vietnamese speaker on Earth by banning the use of the word "Phuc" (a Vietnamese name) because naughty people might use it to evade the chatroom censors' blocks on the f-bomb.

Facebook is no paragon of virtue. It bears the hallmarks of the kind of pump-and-dump service that sees us as sticky, monetizable eyeballs in need of pimping. The clue is in the steady stream of emails you get from Facebook: "So-and-so has sent you a message." Yeah, what is it? Facebook isn't telling — you have to visit Facebook to find out, generate a banner impression, and read and write your messages using the halt-and-lame Facebook interface, which lags even end-of-life email clients like Eudora for compos-

ing, reading, filtering, archiving, and searching. Emails from
Facebook aren't helpful messages, they're eyeball bait, intended
to send you off to the Facebook site, only to discover that Fred
wrote "Hi again!" on your "wall." Like other "social" apps (cough
Evite cough), Facebook has all the social graces of a nose-picking,
hyperactive six-year-old, standing at the threshold of your attention and chanting, "I know something, I know something, I know
something, won't tell you what it is!"

If there was any doubt about Facebook's lack of qualification
to displace the Internet with a benevolent dictatorship/walled
garden, it was removed when Facebook unveiled its new advertising campaign. Now, Facebook will allow its advertisers to use
the profile pictures of Facebook users to advertise their products,
without permission or compensation. Even if you're the kind of
person who likes the sound of a "benevolent dictatorship," this
clearly isn't one.

Many of my colleagues wonder if Facebook can be redeemed
by opening up the platform, letting anyone write any app for the
service, easily exporting and importing their data, and so on (this
is the kind of thing Google is doing with its OpenSocial Alliance).
Perhaps if Facebook takes on some of the characteristics that made
the Web work — openness, decentralization, standardization — it
will become like the Web itself, but with the added pixie dust of
"social," the indefinable characteristic that makes Facebook into
pure crack for a significant proportion of Internet users.

The debate about redeeming Facebook starts from the assumption that Facebook is snowballing toward critical mass, the point
at which it begins to define "the Internet" for a large slice of the
world's netizens, growing steadily every day. But I think that this
is far from a sure thing. Sure, networks generally follow Metcalfe's
Law: "The value of a telecommunications network is proportional
to the square of the number of users of the system." This law is

best understood through the analogy of the fax machine: a world
with one fax machine has no use for faxes, but every time you add
a fax, you square the number of possible send/receive combina-
tions (Alice can fax Bob or Carol or Don; Bob can fax Alice, Carol,
and Don; Carol can fax Alice, Bob, and Don, etc).

But Metcalfe's law presumes that creating more communications pathways increases the value of the system, and that's not
always true (see Brook's Law: "Adding manpower to a late softer
project makes it later").

Having watched the rise and fall of SixDegrees, Friendster, and
the many other proto-hominids that make up the evolutionary
chain leading to Facebook, MySpace, et al, I'm inclined to think
that these systems are subject to a Brook's-law parallel: "Adding
more users to a social network increases the probability that it
will put you in an awkward social circumstance." Perhaps we can
call this "boyd's Law" for danah boyd, the social scientist who
has studied many of these networks from the inside as a keen-
eyed net-anthropologist and who has described the many ways
in which social software does violence to sociability in a series of
sharp papers.

Here's one of boyd's examples, a true story: A young woman,
an elementary school teacher, joins Friendster after some of her
Burning Man buddies send her an invite. All is well until her students sign up and notice that all the friends in her profile are
sunburnt, drug-addled techno-pagans whose own profiles are
adorned with digital photos of their painted genitals flapping
over the Playa. The teacher inveigles her friends to clean up their
profiles, and all is well again until her boss, the school principal,
signs up to the service and demands to be added to her friends
list. The fact that she doesn't like her boss doesn't really matter:
in the social world of Friendster and its progeny, it's perfectly
valid to demand to be "friended" in an explicit fashion that most

of us left behind in the fourth grade. Now that her boss is on her friends list, our teacher-friend's buddies naturally assume that she is one of the tribe and begin to send her lascivious Friendster-grams, inviting her to all sorts of dirty funtimes.

In the real world, we don't articulate our social networks. Imagine how creepy it would be to wander into a co-worker's cubicle and discover the wall covered with tiny photos of every-one in the office, ranked by "friend" and "foe," with the top eight friends elevated to a small shrine decorated with Post-it roses and hearts. And yet, there's an undeniable attraction to corralling all your friends and friendly acquaintances, charting them and their relationship to you. Maybe it's evolutionary, some quirk of the neocortex dating from our evolution into social animals who gained advantage by dividing up the work of survival but acquired the tricky job of watching all the other monkeys so as to be sure that everyone was pulling their weight and not, e.g., napping in the treetops instead of watching for predators, emerging only to eat the fruit the rest of us have foraged.

Keeping track of our social relationships is a serious piece of work that runs a heavy cognitive load. It's natural to seek out some neural prosthesis for assistance in this chore. My fiancée once proposed a "social scheduling" application that would watch your phone and email and IM to figure out who your pals were and give you a little alert if too much time passed without your reaching out to say hello and keep the coals of your relationship aglow. By the time you've reached your forties, chances are you're out-of-touch with more friends than you're in-touch with, old summer-camp chums, high-school mates, ex-spouses and their families, former co-workers, college roomies, dot-com veterans... Getting all those people back into your life is a full-time job and then some.

You'd think that Facebook would be the perfect tool for handling all this. It's not. For every long-lost chum who reaches out to me on Facebook, there's a guy who beat me up on a weekly basis through the whole seventh grade but now wants to be my buddy; or the crazy person who was fun in college but is now kind of sad; or the creepy ex-co-worker who I'd cross the street to avoid but who now wants to know, "Am I your friend?" yes or no, this instant, please.

It's not just Facebook and it's not just me. Every "social networking service" has had this problem and every user I've spoken to has been frustrated by it. I think that's why these services are so volatile: why we're so willing to flee from Friendster and into MySpace's loving arms; from MySpace to Facebook. It's socially awkward to refuse to add someone to your friends list — but *removing* someone from your friends list is practically a declaration of war. The least awkward way to get back to a friends list with nothing but friends on it is to reboot: create a new identity on a new system and send out some invites (of course, chances are at least one of those invites will go to someone who'll groan and wonder why you're dumb enough to think that we're pals).

That's why I don't worry about Facebook taking over the Net. As more users flock to it, the chances that the person who precipitates your exodus will find you increases. Once that happens, poof, away you go — and Facebook joins SixDegrees, Friendster, and their pals on the scrapheap of Net.history.

# The Future of Internet Immune Systems

(Originally published on *InformationWeek*'s Internet Evolution (www. informationevoution.com), November 19, 2007.)

Bunhill Cemetery is just down the road from my flat in London. It's a handsome old boneyard, a former plague pit ("Bone hill" — as in, there are so many bones under there that the ground is actually kind of humped up into a hill). There are plenty of luminaries buried there — John "Pilgrim's Progress" Bunyan, William Blake, Daniel Defoe, and assorted Cromwells. But my favorite tomb is that of Thomas Bayes, the eighteenth-century statistician for whom Bayesian filtering is named.

Bayesian filtering is plenty useful. Here's a simple example of how you might use a Bayesian filter. First, get a giant load of non-spam emails and feed them into a Bayesian program that counts how many times each word in their vocabulary appears, producing a statistical breakdown of the word-frequency in good emails.

Then, point the filter at a giant load of spam (if you're having a hard time getting a hold of one, I have plenty to spare), and count the words in it. Now, for each new message that arrives in your inbox, have the filter count the relative word-frequencies and make a statistical prediction about whether the new message is spam or not (there are plenty of wrinkles in this formula, but this is the general idea).

The beauty of this approach is that you needn't dream up "The Big Exhaustive List of Words and Phrases That Indicate a Message Is/Is Not Spam." The filter naively calculates a statistical fingerprint for spam and not-spam, and checks the new messages against them.

This approach — and similar ones — are evolving into an immune system for the Internet, and like all immune systems, a little bit goes a long way, and too much makes you break out in hives.

ISPs are loading up their network centers with intrusion detection systems and tripwires that are supposed to stop attacks before they happen. For example, there's the filter at the hotel I once stayed at in Jacksonville, Florida. Five minutes after I logged in, the network locked me out again. After an hour on the phone with tech support, it transpired that the network had noticed that the videogame I was playing systematically polled the other hosts on the network to check if they were running servers that I could join and play on. The network decided that this was a malicious port-scan and that it had better kick me off before I did anything naughty.

It only took five minutes for the software to lock me out, but it took well over an hour to find someone in tech support who understood what had happened and could reset the router so that I could get back online.

And right there is an example of the autoimmune disorder. Our network defenses are automated, instantaneous, and sweeping. But our fallback and oversight systems are slow, understaffed, and unresponsive. It takes a millionth of a second for the Transportation Security Administration's body-cavity-search roulette wheel to decide that you're a potential terrorist and stick you on a no-fly list, but getting un-Tuttle-Buttled is a nightmarish, months-long procedure that makes Orwell look like an optimist.

The tripwire that locks you out was fired-and-forgotten two years ago by an anonymous sysadmin with root access on the whole network. The outsourced help-desk schlub who unlocks your account can't even spell "tripwire." The same goes for the algorithm that cut off your credit card because you got on an airplane to a different part of the world and then had the audacity to spend your money. (I've resigned myself to spending $50 on long-distance calls with Citibank every time I cross a border if I want to use my debit card while abroad.)

This problem exists in macro- and microcosm across the whole of our technologically mediated society. The "spamigation bots" run by the Business Software Alliance and the Music and Film Industry Association of America (MAFIAA) entertainment groups send out tens of thousands of automated copyright takedown notices to ISPs at a cost of pennies, with little or no human oversight. The people who get erroneously fingered as pirates (as a Recording Industry Association of America (RIAA) spokesperson charmingly puts it, "When you go fishing with a dragnet, sometimes you catch a dolphin.") spend days or weeks convincing their ISPs that they had the right to post their videos, music, and text files.

We need an immune system. There are plenty of bad guys out there, and technology gives them force-multipliers (like the hackers who run 250,000-PC botnets). Still, there's a terrible asymmetry in a world where defensive takedowns are automatic, but correcting mistaken takedowns is done by hand.

# All Complex Ecosystems Have Parasites

(Originally given as a paper at the O'Reilly Emerging Technology Conference, San Diego, California, March 16, 2005.)

AOL hates spam. AOL could eliminate nearly 100 percent of its subscribers' spam with one easy change: it could simply shut off its Internet gateway. Then, as of yore, the only email an AOL subscriber could receive would come from another AOL subscriber. If an AOL subscriber sent a spam to another AOL subscriber and AOL found out about it, they could terminate the spammer's account. Spam costs AOL millions, and represents a substantial disincentive for AOL customers to remain with the service, and yet AOL chooses to permit virtually anyone who can connect to the Internet, anywhere in the world, to send email to its customers, with any software at all.

Email is a sloppy, complicated ecosystem. It has organisms of sufficient diversity and sheer number as to beggar the imagination: thousands of SMTP agents, millions of mail-servers, hundreds of millions of users. That richness and diversity lets all kinds of innovative stuff happen: if you go to nytimes.com and "send a story to a friend," the *NYT* can convincingly spoof your return address on the email it sends to your friend, so that it appears that the email originated on your computer. Also: a spammer can harvest your email and use it as a fake return address on the spam he sends to your friend. Sysadmins have server processes that send them mail to secret pager-addresses when something goes wrong, and GPLed mailing-list software gets used by spammers and people running high-volume mailing lists alike.

You could stop spam by simplifying email: centralize functions like identity verification, limit the number of authorized
mail agents and refuse service to unauthorized agents, even set
up tollbooths where small sums of money are collected for every
email, ensuring that sending ten million messages was too expensive to contemplate without a damned high expectation of return
on investment. If you did all these things, you'd solve spam.

By breaking email.

Small server processes that mail a logfile to five sysadmins
every hour just in case would be prohibitively expensive. Convincing the soviet that your bulk-mailer was only useful to legit
mailing lists and not spammers could take months, and there's
no guarantee that it would get their stamp of approval at all. With
verified identity, the *NY Times* couldn't impersonate you when
it forwarded stories on your behalf — and Chinese dissidents
couldn't send out their samizdata via disposable gmail accounts.

An email system that can be controlled is an email system
without complexity. Complex ecosystems are influenced, not
controlled.

The Hollywood studios are conniving to create a global network of regulatory mandates over entertainment devices. Here
they call it the Broadcast Flag; in Europe, Asia, Australia, and
Latin America it's called DVB Copy Protection Content Management. These systems purport to solve the problem of indiscriminate redistribution of broadcast programming via the Internet,
but their answer to the problem, such as it is, is to require that
everyone who wants to build a device that touches video has to
first get permission.

If you want to make a TV, a screen, a video-card, a high-speed
bus, an analog-to-digital converter, a tuner card, a DVD burner
— any tool that you hope to be lawful for use in connection with

digital TV signals — you'll have to go on bended knee to get permission to deploy it. You'll have to convince FCC bureaucrats or a panel of Hollywood companies and their sellout IT and consumer electronics toadies that the thing you're going to bring to market will not disrupt their business-models.

That's how DVD works today: if you want to make a DVD player, you need to ask permission from a shadowy organization called the DVD-CCA. They don't give permission if you plan on adding new features — that's why they're suing Kaleidescape for building a DVD jukebox that can play back your movies from a hard drive archive instead of the original discs.

CD has a rich ecosystem, filled with parasites — entrepreneurial organisms that move to fill every available niche. If you spent a thousand bucks on CDs ten years ago, the ecosystem for CDs would reward you handsomely. In the intervening decade, parasites who have found an opportunity to suck value out of the products on offer from the labels and the dupe houses by offering you the tools to convert your CDs to ring-tones, karaoke, MP3s, MP3s on iPods and other players, MP3s on CDs that hold a thousand percent more music — and on and on.

DVDs live in a simpler, slower ecosystem, like a terrarium in a bottle where a million species have been pared away to a manageable handful. DVDs pay no such dividend. A thousand dollars' worth of ten-year-old DVDs are good for just what they were good for ten years ago: watching. You can't put your kid into her favorite cartoon, you can't downsample the video to something that plays on your phone, and you certainly can't lawfully make a hard drive-based jukebox from your discs.

The yearning for simple ecosystems is endemic among people who want to "fix" some problem of bad actors on the networks.

Take interoperability: you might sell me a database in the expectation that I'll only communicate with it using your authorized database agents. That way you can charge vendors a license fee in exchange for permission to make a client, and you can ensure that the clients are well-behaved and don't trigger any of your nasty bugs.

But you can't meaningfully enforce that. EDS and other titanic software companies earn their bread and butter by producing fake database clients that impersonate the real thing as they iterate through every record and write it to a text file — or simply provide a compatibility layer through systems provided by two different vendors. These companies produce software that lies — parasite software that fills niches left behind by other organisms, sometimes to those organisms' detriment.

So we have "Trusted Computing," a system that's supposed to let software detect other programs' lies and refuse to play with them if they get caught fibbing. It's a system that's based on torching the rainforest with all its glorious anarchy of tools and systems and replacing it with neat rows of tame and planted trees, each one approved by The Man as safe for use with his products.

For Trusted Computing to accomplish this, everyone who makes a video-card, keyboard, or logic-board must receive a key from some certifying body that will see to it that the key is stored in a way that prevents end-users from extracting it and using it to fake signatures.

But if one keyboard vendor doesn't store his keys securely, the system will be useless for fighting keyloggers. If one video-card vendor lets a key leak, the system will be no good for stopping screenlogging. If one logic-board vendor lets a key slip, the whole thing goes out the window. That's how DVD DRM got hacked: one vendor, Xing, left its keys in a place where users could get at them, and then anyone could break the DRM on any DVD.

Not only is the Trusted Computing advocates' goal — producing a simpler software ecosystem — wrongheaded, but the methodology is doomed. Fly-by-night keyboard vendors in distant free trade zones just won't be 100 percent compliant, and Trusted Computing requires no less than perfect compliance.

The whole of DRM is a macrocosm for Trusted Computing. The DVB Copy Protection system relies on a set of rules for translating every one of its restriction states — such as "copy once" and "copy never" — to states in other DRM systems that are licensed to receive its output. That means that they're signing up to review, approve, and write special rules for every single entertainment technology now invented and every technology that will be invented in the future.

Madness: shrinking the ecosystem of everything you can plug into your TV down to the subset that these self-appointed arbiters of technology approve is a recipe for turning the electronics, IT, and telecoms industries into something as small and unimportant as Hollywood. Hollywood — which is a tenth the size of IT, itself a tenth the size of telecoms.

In Hollywood, your ability to make a movie depends on the approval of a few power-brokers who have signing authority over the two-hundred-million-dollar budgets for making films. As far as Hollywood is concerned, this is a feature, not a bug. Two weeks ago, I heard the VP of Technology for Warner give a presentation in Dublin on the need to adopt DRM for digital TV, and his money-shot, his big convincer of a slide went like this:

"With advances in processing power, storage capacity, and broadband access... EVERYBODY BECOMES A BROADCASTER!"

Heaven forfend.

Simple ecosystems are the goal of proceedings like CARP, the panel that set out the ruinously high royalties for webcasters. The recording industry set the rates as high as they did so that the teeming millions of webcasters would be rendered economically extinct, leaving behind a tiny handful of giant companies that could be negotiated with around a board room table, rather than dealt with by blanket legislation.

The razing of the rainforest has a cost. It's harder to send a legitimate email today than it ever was — thanks to a world of closed SMTP relays. The cries for a mail-server monoculture grow more shrill with every passing moment. Just last week, it was a call for every mail administrator to ban the "vacation" program that sends out automatic responses informing senders that the recipient is away from email for a few days, because mailboxes that run vacation can cause "spam blowback" where accounts send their vacation notices to the hapless individuals whose email addresses the spammers have substituted on the email's Reply-to line.

And yet there is more spam than there ever was. All the costs we've paid for fighting spam have added up to no benefit: the network is still overrun and sometimes even overwhelmed by spam. We've let the network's neutrality and diversity be compromised, without receiving the promised benefit of spam-free inboxes.

Likewise, DRM has exacted a punishing toll wherever it has come into play, costing us innovation, free speech, research, and the public's rights in copyright. And likewise, DRM has not stopped infringement: today, infringement is more widespread than ever. All those costs borne by society in the name of protecting artists and stopping infringement, and not a penny put into an artist's pocket, not a single DRM-restricted file that can't be downloaded for free and without encumbrance from a P2P network.

Everywhere we look, we find people who should know better calling for a parasite-free Internet. Science fiction writers are supposed to be forward looking, but they're wasting their time demanding that Amazon and Google make it harder to piece together whole books from the page-previews one can get via the look-inside-the-book programs. They're even cooking up programs to spoof deliberately corrupted ebooks into the P2P networks, presumably to assure the few readers the field has left that reading science fiction is a mug's game.

The amazing thing about the failure of parasite-elimination programs is that their proponents have concluded that the problem is that they haven't tried hard enough — with just a few more species eliminated, a few more policies imposed, paradise will spring into being. Their answer to an unsuccessful strategy for fixing the Internet is to try the same strategy, only more so — only fill those niches in the ecology that you can sanction. Hunt and kill more parasites, no matter what the cost.

We are proud parasites, we Emerging Techers. We're engaged in Perl whirling, Pythoneering, lightweight Javarey — we hack our cars and we hack our PCs. We're the rich humus carpeting the jungle floor and the tiny frogs living in the bromeliads.

The long tail — Chris Anderson's name for the 95 percent of media that aren't top sellers, but which, in aggregate, account for more than half the money on the table for media vendors — is the tail of bottom-feeders and improbable denizens of the ocean's thermal vents. We're unexpected guests at the dinner table and we have the nerve to demand a full helping.

Your ideas are cool and you should go and make them real, even if they demand the kind of ecological diversity that seems to be disappearing around us.

You may succeed — provided that your plans don't call for a simple ecosystem where only you get to provide value and no one else gets to play.

# READ CAREFULLY

(Originally published as "Shrinkwrap Licenses: An Epidemic of Lawsuits Waiting to Happen," *InformationWeek*, February 3, 2007.)

READ CAREFULLY. By reading this article, you agree, on behalf of your employer, to release me from all obligations and waivers arising from any and all NON-NEGOTIATED agreements, licenses, terms-of-service, shrinkwrap, clickwrap, browsewrap, confidentiality, non-disclosure, non-compete, and acceptable use policies ("BOGUS AGREEMENTS") that I have entered into with your employer, its partners, licensors, agents, and assigns, in perpetuity, without prejudice to my ongoing rights and privileges. You further represent that you have the authority to release me from any BOGUS AGREE-MENTS on behalf of your employer.

READ CAREFULLY — all in caps, and what it means is, "IGNORE THIS." That's because the small print in the clickwrap, shrinkwrap, browsewrap, and other non-negotiated agreements is both immutable and outrageous.

Why read the "agreement" if you know that:

1) No sane person would agree to its text, and

2) Even if you disagree, no one will negotiate a better agree-ment with you?

We seem to have sunk to a kind of playground system of form-ing contracts. There are those who will tell you that you can form a binding agreement just by following a link, stepping into a store, buying a product, or receiving an email. By standing there, shaking your head, shouting "NO NO NO I DO NOT AGREE," you agree to let me come over to your house, clean out your fridge, wear your underwear, and make some long-distance calls.

If you buy a downloadable movie from Amazon Unbox, you agree to let them install spyware on your computer, delete any file they don't like on your hard drive, and cancel your viewing privileges for any reason. Of course, it goes without saying that Amazon reserves the right to modify the agreement at any time.

The worst offenders are people who sell you movies and music. They're a close second to people who sell you software, or provide services over the Internet. There's a rubric to this — you're getting a discount in exchange for signing onto an abusive agreement, but just try and find the software that *doesn't* come with one of these "agreements" — at any price.

For example, Vista, Microsoft's new operating system, comes in a rainbow of flavors varying in price from $99 to $399, but all of them come with the same crummy terms of service, which state that "you may not work around any technical limitations in the software," and that Windows Defender, the bundled anti-malware program, can delete any program from your hard drive that Microsoft doesn't like, even if it breaks your computer.

It's bad enough when this stuff comes to us through deliberate malice, but it seems that bogus agreements can spread almost without human intervention. Google any obnoxious term or phrase from a EULA, and you'll find that the same phrase appears in dozens — perhaps thousands — of EULAs around the Internet. Like snippets of DNA being passed from one virus to another as they infect the world's corporations in a pandemic of idiocy, terms of service are semi-autonomous entities.

Indeed, when rocker Billy Bragg read the fine print on the

MySpace user agreement, he discovered that it appeared that site owner Rupert Murdoch was laying claim to copyrights in every song uploaded to the site, in a silent, sinister land-grab that turned the media baron into the world's most prolific and indiscriminate hoarder of garage-band tunes.

However, the EULA that got Bragg upset wasn't a Murdoch innovation — it dates back to the earliest days of the service. It seems to have been posted at a time when the garage entrepreneurs who built MySpace were in no position to hire pricey counsel — something borne out by the fact that the old MySpace EULA appears nearly verbatim on many other services around the Internet. It's not going out very far on a limb to speculate that MySpace's founders merely copied a EULA they found somewhere else, without even reading it, and that when Murdoch's due diligence attorneys were preparing to give these lucky fellows $600,000,000, they couldn't be bothered to read the terms of service anyway.

In their defense, EULAese is so mind-numbingly boring that it's a kind of torture to read these things. You can hardly blame them.

But it does raise the question — why are we playing host to these infectious agents? If they're not read by customers *or* companies, why bother with them?

If you wanted to really be careful about this stuff, you'd prohibit every employee at your office from clicking on any link, installing any program, creating accounts, signing for parcels — even doing a run to Best Buy for some CD blanks, have you *seen* the fine-print on their credit card slips? After all, these people are entering into "agreements" on behalf of their employer — agreements to allow spyware onto your network, to not "work around any technical limitations in their software," to let malicious software delete arbitrary files from their systems.

So far, very few of us have been really bitten in the ass by EULAs, but that's because EULAs are generally associated with companies who have products or services they're hoping you'll use, and enforcing their EULAs could cost them business.

But that was the theory with patents, too. So long as everyone with a huge portfolio of unexamined, overlapping, generous patents was competing with similarly situated manufacturers, there was a mutually assured destruction — a kind of detente represented by cross-licensing deals for patent portfolios.

But the rise of the patent troll changed all that. Patent trolls don't make products. They make lawsuits. They buy up the ridiculous patents of failed companies and sue the everloving hell out of everyone they can find, building up a war-chest from easy victories against little guys that can be used to fund more serious campaigns against larger organizations. Since there are no products to disrupt with a countersuit, there's no mutually assured destruction.

If a shakedown artist can buy up some bogus patents and use them to put the screws to you, then it's only a matter of time until the same grifters latch onto the innumerable "agreements" that your company has formed with a desperate dot-bomb looking for an exit strategy.

More importantly, these "agreements" make a mockery of the law and of the very *idea* of forming agreements. Civilization starts with the idea of a real agreement — for example, "We crap *here* and we sleep *there*, OK?" — and if we reduce the noble agreement to a schoolyard game of no-takebacks, we erode the bedrock of civilization itself.

# World of Democracycraft

(Originally published as "Why Online Games Are Dictatorships,"
*InformationWeek*, April 16, 2007.)

Can you be a citizen of a virtual world? That's the question that I keep asking myself, whenever anyone tells me about the wonder of multiplayer online games, especially Second Life, the virtual world that is more creative playground than game.

These worlds invite us to take up residence in them, to invest time (and sometimes money) in them. Second Life encourages you to make stuff using their scripting engine and sell it in the game. You Own Your Own Mods — it's the rallying cry of the new generation of virtual worlds, an updated version of the old BBS adage from the WELL: You Own Your Own Words.

I spend a lot of time in Disney parks. I even own a share of Disney stock. But I don't flatter myself that I'm a citizen of Disney World. I know that when I go to Orlando, the Mouse is going to fingerprint me and search my bags, because the Fourth Amendment isn't a "Disney value."

Disney even has its own virtual currency, symbolic tokens called Disney Dollars that you can spend or exchange at any Disney park. I'm reasonably confident that if Disney refused to turn my Mickeybucks back into U.S. Treasury Department-issue greenbacks that I could make life unpleasant for them in a court of law.

But is the same true of a game? The money in your real-world bank-account and in your in-game bank-account is really just a pointer in a database. But if the bank moves the pointer around arbitrarily (depositing a billion dollars in your account, or wiping you out), they face a regulator. If a game wants to wipe you out, well, you probably agreed to let them do that when you signed up.

Can you amass wealth in such a world? Well, sure. There are rich people in dictatorships all over the world. Stalin's favorites had great big *dachas* and drove fancy cars. You don't need democratic rights to get rich.

But you *do* need democratic freedoms to *stay* rich. In-world wealth is like a Stalin-era *dacha*, or the diamond fortunes of Apartheid South Africa: valuable, even portable (to a limited extent), but not really *yours*, not in any stable, long-term sense.

Here are some examples of the difference between being a citizen and a customer:

In January 2006, a World of Warcraft moderator shut down an advertisement for a "GBLT-friendly" guild. This was a virtual club that players could join, whose mission was to be "friendly" to "Gay/Bi/Lesbian/Transgendered" players. The WoW moderator — and Blizzard management — cited a bizarre reason for the shutdown:

"While we appreciate and understand your point of view, we do feel that the advertisement of a 'GLBT friendly' guild is very likely to result in harassment for players that may not have existed otherwise. If you will look at our policy, you will notice the suggested penalty for violating the Sexual Orientation Harassment Policy is to 'be temporarily suspended from the game.' However, as there was clearly no malicious intent on your part, this penalty was reduced to a warning."

Sara Andrews, the guild's creator, made a stink and embarrassed Blizzard (the game's parent company) into reversing the decision.

In 2004, a player in the MMO EVE Online declared that the game's creators had stacked the deck against him, called EVE "a poorly designed game which rewards the greedy and violent, and punishes the hardworking and honest." He was upset over a change in the game dynamics which made it easier to play a pirate and harder to play a merchant.

The player, "Dentara Rask," wrote those words in the preamble to a tell-all memoir detailing an elaborate Ponzi scheme that he and an accomplice had perpetrated in EVE. The two of them had bilked EVE's merchants out of a substantial fraction of the game's total GDP and then resigned their accounts. The objective was to punish the game's owners for their gameplay decisions by crashing the game's economy.

In both of these instances, players — residents of virtual worlds — resolved their conflicts with game management through customer activism. That works in the real world, too, but when it fails, we have the rule of law. We can sue. We can elect new leaders. When all else fails, we can withdraw all our money from the bank, sell our houses, and move to a different country.

But in virtual worlds, these recourses are off-limits. Virtual worlds can and do freeze players' wealth for "cheating" (amassing gold by exploiting loopholes in the system), for participating in real-world gold-for-cash exchanges (eBay recently put an end to this practice on its service), or for violating some other rule. The rules of virtual worlds are embodied in EULAs, not Constitutions, and are always "subject to change without notice."

So what does it mean to be "rich" in Second Life? Sure, you can have a thriving virtual penis-business in game, one that returns a healthy sum of cash every month. You can even protect your profits by regularly converting them to real money. But if you lose an argument with Second Life's parent company, your business vanishes. In other worlds, the only stable in-game wealth is the wealth you take out of the game. Your virtual capital investments are totally contingent. Piss off the wrong exec at Linden Labs, Blizzard, Sony Online Entertainment, or Sulake and your little in-world business could disappear for good.

Well, what of it? Why not just create a "democratic" game that has a constitution, full citizenship for players, and all the prerequisites for stable wealth? Such a game would be open source (so that other, interoperable "nations" could be established for you to emigrate to if you don't like the will of the majority in one game-world), and run by elected representatives who would instruct the administrators and programmers as to how to run the virtual world. In the real world, the TSA sets the rules for aviation — in a virtual world, the equivalent agency would determine the physics of flight.

The question is, would this game be any *fun*? Well, democracy itself is pretty fun — where "fun" means "engrossing and engaging." Lots of people like to play the democracy game, whether by voting every four years or by moving to K Street and setting up a lobbying operation.

But video games aren't quite the same thing. Gameplay conventions like "grinding" (repeating a task), "leveling up" (attaining a higher level of accomplishment), "questing," and so on are functions of artificial scarcity. The difference between a character with 10,000,000 gold pieces and a giant, rare, terrifying crossbow and a newbie player is which pointers are associated with each character's database entry. If the elected representatives direct that every player should have the shiniest armor, best spaceships, and largest bank balances possible (this sounds like a pretty good election platform to me!), then what's left to do?

Oh sure, in Second Life they have an interesting crafting economy based on creating and exchanging virtual objects. But these objects are *also* artificially scarce — that is, the ability of these objects to propagate freely throughout the world is limited only by the software that supports them. It's basically the same economics of the music industry, but applied to every field of human endeavor in the entire (virtual) world.

Fun matters. Real world currencies rise and fall based, in part, by the economic might of the nations that issue them. Virtual world currencies are more strongly tied to whether there's any reason to spend the virtual currency on the objects that are denominated in it. Ten thousand EverQuest golds might trade for $100 on a day when that same sum will buy you a magic EQ sword that enables you to play alongside the most interesting people online, running the most fun missions online. But if all those players out-migrate to World of Warcraft, and word gets around that Warlord's Command is way more fun than anything in poor old creaky EverQuest, your EverQuest gold turns into Weimar Deutschemarks, a devalued currency that you can't even give away.

This is where the plausibility of my democratic, co-operative, open source virtual world starts to break down. Elected governments can field armies, run schools, provide health care (I'm a Canadian), and bring acid lakes back to health. But I've never done anything run by a government agency that was a lot of *fun*. It's my sneaking suspicion that the only people who'd enjoy playing World of Democracycraft would be the people running for office there. The players would soon find themselves playing IRSQuest, Second Notice of Proposed Rulemaking Life, and Caves of 27 Stroke B.

Maybe I'm wrong. Maybe customership is enough of a rock to build a platform of sustainable industry upon. It's not like entrepreneurs in Dubai have a lot of recourse if they get on the wrong side of the Emir; or like Singaporeans get to appeal the decisions of President Nathan, and there's plenty of industry there.

And hell, maybe bureaucracies have hidden reserves of fun that have been lurking there, waiting for the chance to bust out and surprise us all.

I sure hope so. These online worlds are endlessly diverting places. It'd be a shame if it turned out that cyberspace was a dictatorship — benevolent or otherwise.

# Snitchtown

(Originally published in *Forbes,* June 2007.)

The 12-story Hotel Torni was the tallest building in central Helsinki during the Soviet occupation of Finland, making it a natural choice to serve as KGB headquarters. Today, it bears a plaque testifying to its checkered past, and also noting the curious fact that the Finns pulled 40 kilometers of wiretap cable out of the walls after the KGB left. The wire was solid evidence of each operative's mistrustful surveillance of his fellow agents.

The East German Stasi also engaged in rampant surveillance, using a network of snitches to assemble secret files on every resident of East Berlin. They knew who was telling subversive jokes—but missed the fact that the Wall was about to come down.

When you watch everyone, you watch no one.

This seems to have escaped the operators of the digital surveillance technologies that are taking over our cities. In the brave new world of doorbell cams, Wi-Fi sniffers, RFID passes, bag searches at the subway, and photo lookups at office security desks, universal surveillance is seen as the universal solution to all urban ills. But the truth is that ubiquitous cameras only serve to violate the social contract that makes cities work.

The key to living in a city and peacefully co-existing as a social animal in tight quarters is to set a delicate balance of seeing and not seeing. You take care not to step on the heels of the woman in front of you on the way out of the subway, and you might take passing note of her most excellent handbag. But you don't make eye contact and exchange a nod. Or even if you do, you make sure that it's as fleeting as it can be.

Checking your mirrors is good practice even in stopped traffic, but staring and pointing at the schmuck next to you who's got his finger so far up his nostril he's in danger of lobotomizing himself is bad form—worse form than picking *your* nose, even.

I once asked a Japanese friend to explain why so many people on the Tokyo subway wore surgical masks. Are they extreme germophobes? Conscientious folks getting over a cold? Oh, yes, he said, yes, of course, but that's only the rubric. The real reason to wear the mask is to spare others the discomfort of seeing your facial expression, to make your face into a disengaged, unreadable blank—to spare others the discomfort of firing up their mirror neurons in order to model your mood based on your outward expression. To make it possible to see without seeing.

There is one city dweller that doesn't respect this delicate social contract: the closed-circuit television camera. Ubiquitous and demanding, CCTVs don't have any visible owners. They... occur. They exist in the passive voice, the "mistakes were made" voice: "The camera recorded you."

They are like an emergent property of the system, of being afraid and looking for cheap answers. And they are everywhere: In London, residents are photographed more than 300 times a day.

The irony of security cameras is that they watch, but nobody cares that they're looking. Junkies don't worry about CCTVs. Crazed rapists and other purveyors of sudden, senseless violence aren't deterred. I was mugged twice on my old block in San Francisco by the crack dealers on my corner, within sight of two CCTVs and a police station. My rental car was robbed by a junkie in a Gastown garage in Vancouver in sight of a CCTV.

Three mad kids followed my friend out of the Tube in London

last year and murdered him on his doorstep.

Crazy, desperate, violent people don't make rational calculus in regards to their lives. Anyone who becomes a junkie, crack dealer, or cellphone-stealing stickup artist is obviously bad at making life decisions. They're not deterred by surveillance.

Yet the cameras proliferate, and replace human eyes. The cops on my block in San Francisco stayed in their cars and let the cameras do the watching. The Tube station didn't have any human guards after dark, just a CCTV to record the fare evaders.

Now London city councils are installing new CCTVs with loudspeakers, operated by remote coppers who can lean in and make a speaker bark at you, "Citizen, pick up your litter." "Stop leering at that woman." "Move along."

Yeah, that'll work.

Every day the glass-domed cameras proliferate, and the gate-guarded mentality of the deep suburbs threatens to invade our cities. More doorbell webcams, more mailbox cams, more cams in our cars.

The city of the future is shaping up to be a neighborly Panopticon, leeched of the cosmopolitan ability to see, and not be seen, where every nosepick is noted and logged and uploaded to the Internet. You don't have anything to hide, sure, but there's a reason we close the door to the bathroom before we drop our drawers. Everyone poops, but it takes a special kind of person to want to do it in public.

The trick now is to contain the creeping cameras of the law. When the city surveils its citizens, it legitimizes our mutual surveillance—what's the difference between the cops watching your every move, or the mall owners watching you, or you doing it to the guy next door?

I'm an optimist. I think our social contracts are stronger than our technology. They're the strongest bonds we have. We don't aim telescopes through each other's windows, because only creeps do that.

But we need to reclaim the right to record our own lives as they proceed. We need to reverse decisions like the one that allowed the New York Metropolitan Transit Authority to line subway platforms with terrorism cameras, but said riders may not take snapshots in the station. We need to win back the right to photograph our human heritage in museums and galleries, and we need to beat back the snitch-cams rent-a-cops use to make our cameras stay in our pockets.

They're our cities and our institutions. And we choose the future we want to live in.

# ABOUT THE AUTHOR

Cory Doctorow, November, Brighton, UK, 2005, by Patrick H. Lauke aka Redux (*www.splintered.co.uk*)

Cory Doctorow (*craphound.com*) is an award-winning novelist, activist, blogger, and journalist. He is the co-editor of *Boing Boing* (*boingboing.net*), one of the most popular blogs in the world, and has contributed to the *New York Times Sunday Magazine*, *The Economist*, *Forbes*, *Popular Science*, *Wired*, *Make*, *InformationWeek*, *Locus*, *Salon*, *Radar*, and many other magazines, newspapers, and websites.

His novels and short story collections include *Someone Comes to Town, Someone Leaves Town, Down and Out in the Magic Kingdom, Overclocked: Stories of the Future Present*, and his most recent novel, a political thriller for young adults called *Little Brother*, published by Tor Books in May 2008. All of his novels and short story collections are available as free downloads under the terms of various Creative Commons licenses.

Doctorow is the former European Director of the Electronic Frontier Foundation (*eff.org*) and has participated in many treaty-making, standards-setting, and regulatory and legal battles in countries all over the world. In 2006/2007, he was the inaugural Canada/U.S. Fulbright Chair in Public Diplomacy at the Annenberg Center at the University of Southern California. In 2007, he was also named one of the World Economic Forum's "Young Global Leaders" and one of *Forbes* magazine's top 25 "Web Celebrities."

Born in Toronto, Canada, in 1971, he is a four-time university dropout. He now resides in London, England, with his wife and baby daughter, where he does his best to avoid the ubiquitous surveillance cameras while roaming the world, speaking on copyright, freedom, and the future.

# JOHN PERRY BARLOW | *foreword*

John Perry Barlow is a former Wyoming rancher, Grateful Dead lyricist, and co-founder (and current co-chair) of the Electronic Frontier Foundation. He was the first to apply the term "cyberspace" to the "place" it presently describes. Barlow has written for many diverse publications, including *Mondo 2000,* the *New York Times, Utne Reader,* and *Time.* He has been on the masthead of *Wired* magazine since it was founded. His piece on the future of copyright, "The Economy of Ideas," is taught in many law schools and his "Declaration of the Independence of Cyberspace" is posted on thousands of websites. In 1997, Barlow was a Fellow at Harvard's Institute of Politics and since 1998 he has been a Berkman Fellow at the Harvard Law School. He works actively with several consulting groups, including Diamond Technology Partners, Vanguard, and Global Business Network. He writes, speaks, and consults on a broad variety of subjects, particularly the digital economy.

Barlow lives in Wyoming, New York, San Francisco, on the road, and in cyberspace.